Elementary Statistics in Social Research

THIRD EDITION

Jack Levin

Northeastern University

1817

HARPER & ROW, PUBLISHERS, New York
Cambridge, Philadelphia, San Francisco,
London, Mexico City, São Paulo, Sydney

Sponsoring editor: Alan McClare
Project editor: Eleanor Castellano
Production manager: Jeanie Berke
Compositor: Progressive Typographers
Printer and binder: R. R. Donnelly & Sons Company
Cover illustration: Jeffrey Lindberg, Freelancenter

Elementary Statistics in Social Research, Third Edition
Copyright © 1983 by Harper & Row, Publishers, Inc.

Library of Congress Cataloging in Publication Data

Levin, Jack, 1941–
   Elementary statistics in social research.

   Includes index.
   1. Social sciences—Statistical methods. 2. Statistics. I. Title.
HA29.L388 1983   519.5'024301   82-15456
ISBN 0-06-044071-6

To Flea, Michael, Bonnie, and Andrea

Contents

Preface

Like earlier versions of this text, *Elementary Statistics in Social Research,* Third Edition, is designed to provide an introduction to statistics for students in sociology (and related fields such as political science, social work, psychology, public administration, and education) who have not had extensive training in mathematics and are taking their first course in statistics. As before, this book does *not* purport to be a comprehensive reference work; nor should it be regarded as an appropriate text for advanced courses in statistical methods. On the contrary, it was written and revised in order to fill a perceived need for a meaningful and understandable treatment of basic statistics. Once again, the detailed step-by-step illustrations of statistical procedures have been located at important points throughout the text, and numerous problems from actual research experiences have been provided.

As in the previous editions, this volume has been organized into three parts: Part I (Chapters 2–5) introduces the student to some useful methods for describing and comparing raw data. Part II (Chapters 6 and 7) serves a transitional purpose because it leads the student from the topic of the normal curve—an important descriptive device—to the first chapter in which the normal curve is employed as a basis for generalizing from samples to populations. Continuing with this decision-making focus, Part III (Chapters 8–12) contains several well-known tests of significance, procedures for obtaining correlation coefficients, and an introduction to regression analysis.

This edition has given me a welcome opportunity to correct errors, update examples, and lengthen certain discussions as requested by instructors who have used earlier editons of the text. I have also added a new discussion of the lambda coefficient.

Several pedagogical aids have been added or improved. The number of end-of-chapter exercises has been increased; important terms have been listed at the end of each chapter; and a glossary of terms has been placed at the end of the text. Finally, Chapter 12 now contains an expanded number of research problems and solutions.

A number of individuals have contributed in important ways to the third edition of this text. I am grateful to James A. Fox, Carol Owen, Michael Smith and my students (especially Vanessa Householder, Manuel Jarmille, Ruth Oshman, Rob Roy, Michael W. Smith, Craig Stein, and S. Subramanian) who alerted me to the presence in the previous editions of several errors and sources of misconception. Special thanks are due to the following individuals for their critical reviews of my revisions: Michael G. Horton, Pensacola Junior College; Pamela S. Cain, Hunter College; and Donald T. Matlock, Southwest Texas State University. I am also grateful to Hal Maggied, Nova University, and Meridy Rachar, The Marywood College MSW program at the Lehigh Valley campus, for their unsolicited comments or suggestions. Finally, I am indebted to the Literary Executive of the late Sir Ronald A. Fisher, F. R. S., to Dr. Frank Yates, F. R. S., and to Longman Group Ltd, London, for permission to reprint Tables III, IV, V, and VI from their book *Statistical Tables for Biological, Agricultural and Medical Research* (6th Edition, 1974).

Jack Levin

Elementary Statistics in Social Research

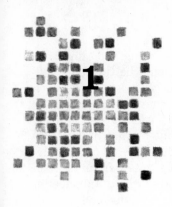

1 Why the Social Researcher Uses Statistics

A LITTLE OF the social scientist can be found in all of us. Almost daily, we take "educated guesses" concerning the future events in our lives in order to predict what will happen in new situations or experiences. As these situations occur, we are sometimes able to confirm or support our ideas; other times, however, we are not so lucky and must experience the sometimes unpleasant consequences.

To take some familiar examples: we might invest in the stock market, vote for a political candidate who promises to solve domestic problems, "play" the horses, take medicine to reduce the discomfort of a cold, throw a pair of dice at a gambling casino, try to "psych out" our instructors regarding a midterm, or accept a "blind date" on the word of a friend.

Sometimes we win; sometimes we lose. Thus, we might make a sound investment in the stock market, but be sorry about our voting decision; win money at the crap table, but discover we have taken the wrong medicine for our illness; do well on a midterm, but have a miserable blind date, and so on. It is unfortunately true that not all of our everyday predictions will be supported by experience.

THE NATURE OF SOCIAL RESEARCH

In a somewhat similar manner, social scientists have ideas about the nature of social reality (which they label *hypotheses*) and frequently test their ideas by doing systematic research. Hypotheses often take the form of a statement of relationship between two variables: an *independent variable* (or presumed cause) and a *dependent variable* (or presumed effect). For example, a researcher might hypothesize that socially isolated children watch more television than children who are well integrated into their peer groups, and he or she might do a survey in which both socially isolated and well-

1

integrated children are asked questions regarding the time they spend watching television (social isolation would be the independent variable; TV viewing behavior would be the dependent variable). Or a researcher might hypothesize that the one-parent family structure (absent mother or father) generates greater delinquency than the two-parent family structure (present mother and father) and might proceed to interview samples of delinquents and nondelinquents to determine whether one or both parents were present in their family backgrounds (family structure would be the independent variable; delinquency would be the dependent variable).

Thus, not unlike their counterparts in the physical sciences, social researchers often conduct research in order to increase their understanding of the problems and issues in their field. Social research takes many forms and can be used to investigate a wide range of problems. The researcher may work on a participant observation of delinquent gangs, a sample survey of political likes and dislikes, an analysis of values in the underground press, or an experiment to determine the effects of forcing families to relocate their homes in order to make room for newly constructed highways.

Among the most useful research methods employed by social researchers for the purpose of testing their hypotheses are the experiment, the survey, content analysis, and participant observation.

The Experiment

Unlike everyday observation (or, for that matter, any other research approach), the *experiment* is distinguished by the degree of *control* a researcher is able to apply to the research situation. In an experiment, researchers actually manipulate one or more of the independent variables to which their subjects are exposed. The manipulation occurs when an experimenter assigns the independent variable to one group of people (called an *experimental group*), but withholds it from another group of people (called a *control group*). Ideally, all other initial differences between the experimental and control groups are eliminated by matching members of both groups on important characteristics (for example, by having the same proportion of men and women in each group) or by assigning subjects on a random basis to the experimental and control conditions.

For example, a researcher who hypothesizes that frustration increases aggression might assign a number of subjects to the experimental and control groups on a random basis by the flip of a coin ("heads" you're in the experimental group; "tails" you're in the control group), so that the groups initially differ only by chance. The researcher might then manipulate frustration (the independent variable) by asking the members of the experimental group to solve a difficult (frustrating) puzzle, while the members of the control group are asked instead to solve a much easier (nonfrustrating) version of the same puzzle. After all subjects have been given a period of time to complete their puzzle, the researcher might obtain a measure of aggression by asking them to administer "a mild electrical shock" to another subject (actually, the other subject is a confederate of the researcher who never really gets shocked, but the naive subject presumably does not know this). If the willingness of subjects to administer an electrical shock is greater in the experimental group than in the control group, this difference would be attributed to the effect of the independent variable, frustration. Conclusion: frustration does indeed tend to increase aggressive behavior.

The Survey

As we have seen, the experimenter actually has a direct hand in creating the effect that he or she seeks to achieve. By contrast, *survey* research is *retrospective*—the effects of independent variables on dependent variables are *recorded* after—and sometimes long after—they have occurred. Survey researchers typically seek to reconstruct these influences and consequences by means of the verbal reports of their respondents, in self-administered questionnaires or in face-to-face interviews.

Surveys lack the controls of experiments: no manipulation occurs. But surveys also have their relative advantages precisely because they do not involve an experimental manipulation. As compared with experiments, survey research can investigate a much larger number of important independent variables in relation to any dependent variable. Because they are not confined to a laboratory setting in which an independent variable can be manipulated, surveys can also be more *representative*—their results can be generalized to a broader range of people.

For example, a survey researcher who hypothesizes that frustration increases aggression might locate a number of severely aggressive individuals who are interviewed in order to identify the frustrating events in their lives such as isolation, physical disabilities, poor grades in school, and poverty. Obviously, survey researchers cannot introduce these frustrating life events themselves; but they can attempt to discover and record them after they have occurred. To study the relationship between frustration and aggression, Stuart Palmer interviewed the mothers of 51 convicted murderers. He found many more frustrating circumstances in the early lives of these killers than in the lives of their "control" brothers who had not committed murder. Specifically, the mothers recalled that their murdering children had been subjected to more serious illnesses, operations, accidents, beatings, physical defects, frightening experiences, and disapproval from their peers.

Content Analysis

As an alternative to experiments and surveys, *content analysis* is that research method whereby a researcher seeks objectively to describe the content of communication messages that people have previously produced. As a result, researchers who conduct a content analysis have no need directly to observe behavior or to question a set of respondents. Instead, they typically study the content of books, magazines, newspapers, films, radio broadcasts, photographs, cartoons, letters, verbal dyadic interaction, political propaganda, or music.

For example, in 1977, Allan Kimmel and I content analyzed syndicated newspaper gossip columns from the time intervals 1954–1955, 1964–1965, and 1974–1975. Using appropriate recording sheets, we coded each instance of gossip occurring in our sample of gossip columns for such characteristics as sex, race, and occupation of the target of gossip. Our results indicated that these celebrities tended to be white, male, and "star" show people such as actors, dancers, or directors. We also found, however, that the representation of show people in these gossip columns decreased significantly from 1954–1955 to 1974–1975 (an increased presence of politicians made up most of the difference over time). Perhaps our most important finding was that much of the behavior of the targets of gossip received explicit approval. Despite their reputation for "spreading dirt," these gossip columnists wrote nice things about the "Stars"—they tended to stress public recognition of approved behavior, while generally ignoring the negative sanction.

Participant Observation

Another widely used research method is *participant observation,* whereby a researcher "participates in the daily life of the people under study, either openly in the role of researcher or covertly in some disguised role, observing things that happen, listening to what is said, and questioning people, over some length of time."

The particular strength of participant observation is its ability to provide a *complete* form of information about a situation or a series of events. The participant observer is able to determine the meaning of social situations, not from the viewpoint of outsiders, but as defined by the group members themselves.

A classic example of participant observation is found in John Dollard's early study of the patterns of race relations in a small southern town. For a five-month period during 1936, Dollard operated in the role of researcher to observe what the townspeople said, did, and seemed to feel, without directly interrogating members of the community with predetermined questions. Based on his participant observation, Dollard discovered, among other things, that white, middle-class residents of this southern town gained some important economic, sexual, and prestige benefits from the maintenance of traditional patterns of race relations between blacks and whites. He also found that race prejudice and discrimination generated forms of aggression among blacks, against whites, and against blacks.

WHY TEST HYPOTHESES?

It is usually desirable, if not necessary, to systematically test our hypotheses about the nature of social reality, even those that seem true, logical, or self-evident. Our everyday commonsensical "tests" are generally based on very narrow, if not biased, preconceptions and personal experiences, which can lead us to accept invalid conclusions regarding the nature of social phenomena. To demonstrate this point, let us examine the following hypotheses, which were tested on a large number of soldiers during World War II. Would you have predicted these results on the basis of your everyday experiences? Do you think they were worthy of testing? Or do they seem too obvious and self-evident for systematic research?

1. Better-educated men showed more psychoneurotic symptoms than those with less education.
2. Men from rural backgrounds were usually in better spirits during their army life than soldiers from city backgrounds.
3. Southern soldiers were better able to stand the climate in the hot South Sea Islands than northern soldiers.
4. As long as the fighting continued, men were more eager to be returned to the States than they were after the German surrender.

If you believe that these relationships were too commonsensical for systematic testing, then you might be interested to learn that every statement "is the direct opposite of what actually was found. Poorly educated soldiers were more neurotic than those with high education; Southerners showed no greater ability than Northerners to adjust to a tropical climate; . . . and so on."[1] To depend on common sense or everyday experience alone *obviously* has its limits!

THE STAGES OF SOCIAL RESEARCH

Systematically testing our ideas about the nature of social reality often demands carefully planned and executed research in which:

1. the problem to be studied is reduced to a testable hypothesis (for example, "one-parent families generate more delinquency than two-parent families");
2. an appropriate set of instruments is developed (for example, a questionnaire or an interview schedule is constructed);
3. data are collected (that is, the researcher might go into the field and do a poll or a survey);
4. the data are analyzed for their bearing on the initial hypotheses; and
5. results of the analysis are interpreted and communicated to an audience, for example, by means of a lecture or journal article.

[1] Paul Lazarsfeld, "The American Soldier—An Expository Review," *Public Opinion Quarterly,* fall, 1949, p. 380.

As we shall see in subsequent chapters, the material presented in this book is most closely tied to the data-analyzing stage of research (see 4 above), in which the data collected or gathered by the researcher are analyzed for their bearing on the initial hypotheses. It is in this stage of research that the raw data are tabulated, calculated, counted, summarized, rearranged, compared or, in a word, *organized,* so that the accuracy or validity of our hypotheses can be tested.

USING SERIES OF NUMBERS TO DO SOCIAL RESEARCH

Anyone who has conducted social research knows that problems in the analysis of data must be confronted in the planning stages of a research project since they have a bearing on the nature of decisions at all other stages. Such problems often affect aspects of the research design and even the types of instruments that are to be employed in collecting data. For this reason, we constantly seek techniques or methods for enhancing the quality of data analysis.

Many researchers feel that it is essential to employ *measurement* or a series of numbers in analyzing data. Therefore, social researchers have developed measures of a wide range of phenomena, including occupational prestige, political attitudes, authoritarianism, alienation, anomie, delinquency, social class, prejudice, dogmatism, conformity, achievement, ethnocentrism, neighborliness, religiosity, marital adjustment, occupational mobility, urbanization, sociometric status, and fertility.

Numbers have at least three important functions for the social researcher, depending on the particular *level of measurement* which he or she employs. Specifically, series of numbers can be used

1. to categorize at the *nominal* level of measurement,
2. to *rank* or *order* at the *ordinal* level of measurement, and
3. to *score* at the *interval* level of measurement.

Before moving on to a discussion of the role of statistics in social research, let us stop here to examine some of the major characteristics of these levels of measurement, characteristics that will later take on considerable meaning when we attempt to apply statistical techniques to particular research situations.

The Nominal Level

The *nominal* level of measurement simply involves the process of naming or labeling; that is, of placing cases into categories and counting their frequency of occurrence. To illustrate, we might use a nominal-level measure to indicate whether each respondent is prejudiced or unprejudiced in his attitude toward Puerto Ricans. As shown in Table 1.1, we might question the 10 students in a given class and determine that 5 can be regarded as (1) prejudiced, and 5 can be considered (2) unprejudiced.

Other nominal-level measures in social research are sex (male

TABLE 1.1
Attitudes toward Puerto Ricans of ten college students: nominal data

Attitude Toward Puerto Ricans	Frequency
1 = prejudiced	5
2 = unprejudiced	5
Total	10

versus female), welfare status (recipient versus nonrecipient), political party (Republican, Democrat, Independent, and Socialist), social character (inner-directed, other-directed, and tradition-directed), mode of adaptation (conformity, innovation, ritualism, retreatism, and rebellion), time orientation (present, past, and future), and urbanization (urban, rural, and suburban), to mention only a few.

When dealing with nominal data, we must keep in mind that *every case must be placed in one, and only one, category.* This requirement indicates that the categories must be nonoverlapping or *mutually exclusive.* Thus, a respondent's race classified as "white" cannot also be classified as "black"; any respondent labeled "male" cannot also be labeled "female." The requirement also indicates that the categories must be *exhaustive*—there must be a place for every case that arises. For illustrative purposes, imagine a study in which all respondents are interviewed and categorized by race as either black or white. Where would we categorize a Chinese respondent, if he were to appear? In this case, it might be necessary to expand the original category system to include "Orientals" or, assuming that most respondents will be white or black, to include a miscellaneous category in which such exceptions can be placed.

The reader should note that nominal data are not graded, ranked, or scaled for qualities such as better or worse, higher or lower, more or less. Clearly, then, a nominal measure of sex does not signify whether males are "superior" or "inferior" to females. Nominal data are merely labeled, sometimes by name (male versus female or prejudiced versus unprejudiced), other times by number (1 versus 2), but always for the purpose of grouping the cases into separate categories to indicate sameness or differentness with respect to a given quality or characteristic.

The Ordinal Level

When the researcher goes beyond this level of measurement and seeks to order his or her cases in terms of the degree they have any given characteristic, he or she is working at the *ordinal* level of measurement. The nature of the relationship among ordinal categories depends on that characteristic which the researcher seeks to measure. To take a familiar example, one might classify individuals with respect to socioeconomic status as "lower class," "middle class," or "upper class." Or, rather than categorize the students in a given classroom as *either* prejudiced *or* unprejudiced, he or she might rank

TABLE 1.2
Attitudes toward Puerto Ricans of ten college students: ordinal data

Student	Rank
Joyce	1—most prejudiced
Paul	2—second
Cathy	3—third
Mike	4—fourth
Judy	5—fifth
Joe	6—sixth
Kelly	7—seventh
Ernie	8—eighth
Linda	9—ninth
Ben	10—least prejudiced

them according to their degree of prejudice against Puerto Ricans as indicated in Table 1.2.

The ordinal level of measurement yields information about the ordering of categories, but does not indicate the *magnitude of differences* between numbers. For instance, the social researcher who employs an ordinal-level measure to study prejudice toward Puerto Ricans *does not know how much more prejudiced one respondent is than another*. In the example given above, it is not possible to determine how much more prejudiced Joyce is than Paul or how much less prejudiced Ben is than Linda or Ernie. This is because the intervals between the points or ranks on an ordinal scale are not known or meaningful. Therefore it is not possible to assign *scores* to cases located at points along the scale.

The Interval Level

By contrast, the *interval* level of measurement tells us about the ordering of categories and also indicates the exact *distance* between them. Interval measures use *constant units of measurement* (for example, dollars or cents, Fahrenheit or Celsius, yards or feet, minutes or seconds), which yield *equal intervals* between points on the scale.

In this way, an interval measure of prejudice against Puerto Ricans—such as responses to a series of questions about Puerto Ricans that are scored from 0 to 100 (100 is extreme prejudice)—might yield the data shown in Table 1.3 about the ten students in a given classroom.

As presented in Table 1.3, we are able to order the students in terms of their prejudices and, in addition, indicate the distances separating one from another. For instance, it is possible to say that Ben is the least prejudiced member of the class since he received the lowest score. We can say also that Ben is only slightly less prejudiced than Linda or Ernie but much less prejudiced than Joyce, Paul, Cathy, or Mike, all of whom received extremely high scores. Depending on the purpose for which the study is designed, such information might be important to determine, but is not available at the ordinal level of measurement.

TABLE 1.3
Attitudes toward Puerto Ricans of ten college students: interval data

Student	Score[a]
Joyce	98
Paul	96
Cathy	95
Mike	94
Judy	22
Joe	21
Kelly	20
Ernie	15
Linda	11
Ben	6

[a] Higher scores indicate greater prejudice against Puerto Ricans.

FUNCTIONS OF STATISTICS

It is when researchers use numbers—they *quantify* their *data* at the nominal, ordinal, or interval level of measurement—that they are likely to employ statistics as a tool of (1) *description* or (2) *decision making*. Let us now take a closer look at these important functions of statistics.

Description

To arrive at conclusions or obtain results, a social researcher often studies hundreds, thousands, or even larger numbers of persons or groups. As an extreme case, the United States Bureau of the Census conducts a complete enumeration of the United States population, in which more than 200 million individuals are contacted. Despite the aid of numerous sophisticated procedures designed for the purpose, it is always a formidable task to describe and summarize the masses of data that are generated from projects in social research.

To take a familiar example, the examination grades of a class of only 80 students have been listed in Table 1.4. Do you see any patterns in these grades? Can you describe these grades in a few words? In a few sentences? Are they particularly high or low on the whole?

Using even the most basic principles of descriptive statistics as presented in subsequent chapters of this text, it is possible to characterize the distribution of examination grades in Table 1.4 with a good deal of clarity and precision, so that overall tendencies or group characteristics can be more quickly discovered and easily communicated to almost anyone. First, we might rearrange the grades in consecutive order (from highest to lowest) in order to group them into a much smaller number of categories. As shown in Table 1.5, this *grouped frequency distribution* (to be discussed in detail in Chapter 2) would present the grades within broader categories along with the number or *frequency* (*f*) of students whose grades fell into these categories.

It can readily be seen, for example, that 17 students received grades between 60 and 69; only 2 students received grades between 20 and 29.

TABLE 1.4
Examination grades for 80 students

72	83	91	29
38	89	49	36
43	60	67	49
81	52	76	62
79	62	72	31
71	32	60	73
65	28	40	40
59	39	58	38
90	49	52	59
83	48	68	60
39	65	54	75
42	72	52	93
58	81	58	53
56	58	77	57
72	45	88	61
63	52	70	65
49	63	61	70
81	73	39	79
56	69	74	37
60	75	68	46

Another useful procedure (explained in Chapter 3) might be to rearrange the grades graphically. As shown in Figure 1.1, we might place the categories of grades (from 20–29 to 90–99) along one line of a graph (that is, the *horizontal base line*) and their numbers or frequencies along another line (that is, the *vertical axis*).

This arrangement results in a rather easily visualized graphic representation (for example, the bar graph), in which we can see that most grades fall between 50 and 80, whereas relatively few grades are much higher or lower.

As will be elaborated in Chapter 4, a particularly convenient and useful statistical method—one with which we are already more or less familiar—is to ask: what is the grade of the *average* person in this group of 80 students? The arithmetic average (or *mean*), which can be obtained by adding the entire list of grades and dividing this sum by the number of students, gives us a clearer picture of the

TABLE 1.5
Examination grades for 80 students: a grouped frequency distribution

Grades	f
90–99	3
80–89	7
70–79	16
60–69	17
50–59	15
40–49	11
30–39	9
20–29	2

FIGURE 1.1
Examination grades for 80 students arranged in a bar graph

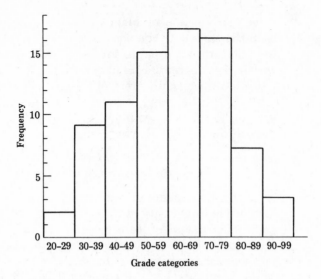

overall group tendency. The arithmetic average in the present illustration happens to be 60.5, a rather low grade as compared against class averages with which most students may already be familiar. Apparently, this group of 80 students did relatively poorly as a whole!

Thus, with the help of statistical devices, such as grouped frequency distributions, graphs, and the arithmetic average, it is possible to detect and describe patterns or tendencies in distributions of scores (for example, the grades located in Table 1.4), which might otherwise have gone unnoticed by the casual observer. In the present context, then, statistics may be defined as a *set of techniques for the reduction of quantitative data (that is, a series of numbers) to a small number of more convenient and easily communicated descriptive terms.*

Decision Making

For purposes of testing hypotheses, it is frequently necessary to go beyond mere description. It is often also necessary to make inferences; that is, to make decisions based on data collected on only a small portion or *sample* of the larger group we have in mind to study. Factors such as cost, time, and need for adequate supervision many times preclude taking a complete enumeration or poll of the entire group (social researchers call this larger group from which the sample was drawn a *population* or *universe*).

As we shall see in Chapter 7, every time the social researcher tests his hypotheses on a sample, he must decide whether it is indeed accurate to generalize his findings to the entire population from which they were drawn. Error inevitably results from sampling, even sampling that has been properly conceived and executed. This is the problem of generalizing or *drawing inferences* from the sample to the population.[2]

Statistics can be useful for purposes of generalizing findings, with a high degree of confidence, from small samples to larger populations. To better understand this decision-making purpose of statistics and the concept of generalizing from samples to populations, let us examine the results of a hypothetical study which was conducted to test the following hypothesis:

Hypothesis: Male college students are more likely than female college students to have tried marijuana.

The researchers in this study decided to test their hypothesis at an urban university in which some 20,000 students (10,000 males and 10,000 females) were enrolled. Due to cost and time factors, they were not able to interview every student on campus, but did obtain from the registrar's office a complete listing of university students. From this list, every one hundredth student (one-half of them male, one-half of them female) was selected for the sample and subsequently interviewed by members of the research staff who had been trained for this purpose. The interviewers asked each of the 200 members of the sample whether he or she had ever tried marijuana and then recorded the student's sex as either male or female. After all interviews had been completed and returned to the staff office, the results of the study were tabulated by sex and presented in Table 1.6.

Notice that results obtained from this sample of 200 students

[2] *To the student:* The concept "sampling error" is discussed in greater detail in Chapter 7. However, to better understand the inevitability of error when sampling from a larger group, the student may now wish to conduct the following demonstration. Refer to Table 1.4, which contains the grades of a population of 80 students. At "random" (for example, by closing your eyes and pointing), select a sample of a few grades (for example, 5) from the entire list. Find the average grade by adding the five scores and dividing by 5, the total number of grades. It has already been pointed out that the average grade for the entire class of 80 students was 60.5. To what extent does your sample average differ from the class average, 60.5? Try this demonstration on several more samples of a few grades randomly selected from the larger group. With great consistency you should find that your sample mean will almost always differ at least slightly from that obtained from the entire class of 80 students. This is what we mean by "sampling error."

TABLE 1.6
Marijuana use by sex of respondent: case I

Marijuana Use	Sex of Respondents	
	Males	*Females*
Number who have tried it	60	40
Number who have not tried it	40	60
Total	100	100

TABLE 1.7
*Marijuana use by sex of
respondent: case II*

	Sex of Respondents	
Marijuana Use	*Males*	*Females*
Number who have tried it	55	45
Number who have not tried it	45	55
Total	100	100

as presented in Table 1.6 are in the hypothesized direction: 60 out of
100 males reported having tried marijuana, whereas only 40 out of
100 females reported having tried marijuana. Clearly, in this small
sample, males were more likely than females to have tried marijuana. For our purposes, however, the more important question is
whether these sex differences in marijuana use are large enough so
that we might confidently generalize them to the much larger university population of 20,000 students. Do these results represent
true population differences? Or have we obtained chance differences
between males and females due strictly to sampling error—the
error that occurs every time we take a small group from a larger
group?

To illuminate the problem of generalizing results from samples
to larger populations, let us imagine that the researchers had, instead, obtained the results shown in Table 1.7. Notice that these results are still in the predicted direction: 55 males as opposed to only
45 females have tried marijuana. But, are we still willing to generalize these results to the larger university population? Is it not likely
that a difference of this magnitude (10 more males than females)
would have happened simply by chance? Or can we confidently say
that such relatively small differences reflect a real difference
between males and females at that particular university?

Let us carry out the illustration a step further. Suppose that
the social researchers had obtained the data shown in Table 1.8. Differences between males and females shown in the table could not be
much smaller and still be in the hypothesized direction: 51 males in
contrast to 49 females tried marijuana—only 2 more males than females. How many of us would be willing to call *this* finding a true
population difference between males and females, rather than a
product of chance or sampling error?

Where do we draw the line? At what point does a sample dif-

TABLE 1.8
*Marijuana use by sex of
Respondent: case III*

	Sex of Respondents	
Marijuana Use	*Males*	*Females*
Number who have tried it	51	49
Number who have not tried it	49	51
Total	100	100

ference become large enough so that we are willing to treat it as significant or real? With the aid of statistics, we can readily, and with a high degree of confidence, make such decisions about the relationship between samples and populations. To illustrate, had we used one of the statistical tests of significance later discussed in this text (for example, chi square—see Chapter 10), we would already have known that *only those results* reported in Table 1.6 can be generalized to the population of 20,000 university students—that 60 out of 100 males but only 40 out of 100 females have tried marijuana is a finding substantial enough to be applied to the entire population with a high degree of confidence. Our statistical test tells us there are only 5 chances out of 100 that we are wrong! By contrast, results reported in Tables 1.7 and 1.8 are statistically *nonsignificant,* being the product of sampling error rather than real sex differences in the use of marijuana. Again, using a statistical criterion, we conclude that these results do not reflect true population differences, but are a mere sampling error.

In the present context, then, statistics is *a set of decision-making techniques which aid researchers in drawing inferences from samples to populations and, hence, in testing hypotheses regarding the nature of social reality.*

SUMMARY

This chapter linked our everyday predictions about future events with the experiences of the social researcher who uses statistics as an aid in testing his hypotheses about the nature of social reality. Measurement was discussed in terms of nominal, ordinal, and interval data. Two major functions of statistics were identified with the data-analysis stage of social research, then briefly discussed and illustrated:

1. description (that is, the reduction of quantitative data to a small number of more convenient descriptive terms) and
2. decision making (that is, drawing inferences from samples to populations).

PROBLEMS

Identify the level of measurement—nominal, ordinal, or interval—represented in each of the following questionnaire items:

1. Your sex: 1 _____ Female
              2 _____ Male

2. Your age: 1 _____ Younger than 13
              2 _____ 13–19
              3 _____ 20–29
              4 _____ 30–39
              5 _____ 40–49
              6 _____ 50–59
              7 _____ 60–69
              8 _____ 70 or older

3. Your IQ: _____
     (specify exact score)
4. Specify the highest level of education achieved by your father:
   1 ___ None
   2 ___ Elementary school
   3 ___ Some high school
   4 ___ Graduated high school
   5 ___ Some college
   6 ___ Graduated college
   7 ___ Graduate school
5. Your annual income from all sources:

   _____
   (specify precise amount)
6. Your religious preference:
   1 ___ Protestant
   2 ___ Catholic
   3 ___ Jewish
   4 ___ Other _____
     (specify)
7. The social class to which your parents belong:
   1 ___ Upper
   2 ___ Upper-middle
   3 ___ Middle-middle
   4 ___ Lower-middle
   5 ___ Lower
8. In which of the following areas do your parents presently live?
   1 ___ Rural
   2 ___ Small town
   3 ___ Suburb near a city
   4 ___ City
9. Indicate your political orientation by placing an "X" in the appropriate space:
   LIBERAL ___:___:___:___:___ CONSERVATIVE
       1   2   3   4   5

TERMS TO
REMEMBER

Hypothesis
Independent and dependent variables
Experiment
Survey
Content analysis
Participant observation
Measurement
Level of measurement
  Nominal
  Ordinal
  Interval
Description and decision-making functions of statistics

PART I
DESCRIPTION

2 Organizing the Data

COLLECTING THE DATA entails a serious effort on the part of social researchers who seek to increase their knowledge of human behavior. To interview or otherwise elicit information from a number of welfare recipients, college students, drug addicts, public housing residents, homosexuals, middle-class Americans, or other respondents requires a degree of foresight, careful planning, and control, if not actual time spent in the field.

The completion of data collection is, however, only the beginning as far as statistical analysis is concerned. Data collection yields the raw materials with which social researchers must work, if they are to analyze data, obtain results, and test hypotheses about the nature of social reality.

FREQUENCY DISTRIBUTIONS OF NOMINAL DATA

The cabinetmaker transforms raw wood into furniture; the chef converts raw food into the more palatable versions served at the dinner table. By a similar process, the social researcher—aided by "recipes" called formulas and techniques—attempts to transform raw data into a meaningful and organized set of measures which can be used to test the initial hypotheses.

What can social researchers do to organize the jumble of raw numbers that they collect from their respondents? How do they go about transforming this mass of raw data into an easy-to-understand summary form? The first step might be to construct a *frequency distribution* in the form of a table.

Let us examine the frequency distribution in Table 2.1. Notice first of all that the table is headed by a *number* (2.1) and a *title* which gives the reader an idea as to the nature of the data being

19

TABLE 2.1
Male and female students majoring in engineering

Sex of Student	Frequency (f)
Male	80
Female	20
Total	100

presented—"Male and Female Students Majoring in Engineering." This is the standard arrangement; every table must be clearly titled and, when presented in a series, labeled by number as well.

Frequency distributions of nominal data consist of two columns. As in Table 2.1, the left-hand column indicates what characteristic is being presented (sex of student) and contains the categories of analysis (male and female). An adjacent column is headed "Frequency" or "f" and indicates the number of cases in each category (80 and 20, respectively) as well as the total number of cases ($N = 100$). A quick glance at the frequency distribution in Table 2.1 clearly reveals that many more males than females majored in engineering—80 out of 100 students majoring in engineering on this hypothetical campus were males.

COMPARING DISTRIBUTIONS

Suppose, however, we wish to compare the engineering majors on two different college campuses. Making comparisons *between* frequency distributions is a procedure often used to clarify results and add information. The particular comparison a researcher makes is determined by the question he or she seeks to answer.

Returning to our hypothetical example, we might ask: are male students more likely than female students to major in engineering on campuses A and B? To provide an answer, we might compare the 100 students majoring in engineering at College A with another 100 students who are majoring in engineering at College B. Let us imagine that the data shown in Table 2.2 are obtained.

As shown in the table, 30 out of 100 engineering majors at College B, but only 20 out of 100 engineering majors at College A, were females. Thus, we can now say that the male dominance among

TABLE 2.2
Male and female students majoring in engineering at colleges A and B

	Engineering Majors	
	College A	College B
Sex of Student	f	f
Male	80	70
Female	20	30
Total	100	100

engineering majors occurs on more than one college campus. We can also say that female students at College B are somewhat more likely to major in engineering than are female students at College A.

PROPORTIONS AND PERCENTAGES

When the researcher studies distributions of equal total size, the frequency data can be used to make comparisons between the groups. Thus, the number of males majoring in engineering on compuses A and B can be directly compared, because there were exactly 100 majors on each campus. It is generally not possible, however, to study distributions having exactly the same number of cases. For instance, how can we make sure that precisely 100 students at both colleges will decide to major in engineering?

In order to clarify such results, we need a method of *standardizing frequency distributions* for size—a way to compare groups despite differences in total frequencies. Two of the most popular and useful methods of standardizing for size and comparing distributions are the *proportion* and the *percentage*.

The proportion compares the number of cases in a given category with the total size of the distribution. We can convert any frequency into a proportion P by dividing the number of cases in any given category f by the total number of cases in the distribution N. Or,

$$P = \frac{f}{N}$$

Therefore, 10 males out of 40 students majoring in engineering can be expressed as the proportion $P = \frac{10}{40} = .25$.

Despite the usefulness of the proportion, many people prefer to indicate the relative size of a series of numbers in terms of the *percentage*, the frequency of occurrence of a category *per 100 cases*. To calculate a percentage, we simply multiply any given proportion by 100. By formula,

$$\% = (100)\frac{f}{N}$$

Therefore, 10 males out of 40 students majoring in engineering can be expressed as the proportion $P = \frac{10}{40} = .25$, or as a percentage $\% = (100)\frac{10}{40} = 25$ percent. Thus, 25 percent of this group of 40 engineering majors are males. To illustrate the utility of percentages in making comparisons between distributions, let us examine the engineering majors at two colleges, one where engineering is a popular major and another where it is not. Let us suppose, for example, that College A had 1352 engineering majors whereas College B had only 183 engineering majors.

Table 2.3 indicates both the frequencies and the percentages for engineering majors at Colleges A and B. Notice the difficulty one

TABLE 2.3
Male and female engineering majors at colleges A and B

| | Engineering Majors | | | |
| | College A | | College B | |
Sex of Student	*f*	*%*	*f*	*%*
Male	1082	(80)	146	(80)
Female	270	(20)	37	(20)
Total	1352	(100)	183	(100)

has in quickly determining sex differences among engineering majors from the frequency data alone. By contrast, the percentages clearly reveal that females were equally represented among engineering majors at Colleges A and B. Specifically, 20 percent of the engineering majors at College A were females; 20 percent of the engineering majors at College B were females.

Ratios

A less commonly used method of standardizing for size, the *ratio*, directly compares the number of cases falling into one category (for example, males) with the number of cases falling into another category (for example, females). Thus, a ratio can be obtained in the following manner where f_1 = frequency in any category, and f_2 = frequency in any other category:

$$\text{ratio} = \frac{f_1}{f_2}$$

If we were interested in determining the ratio of blacks to whites, we might compare the number of black respondents ($f = 150$) to the number of white respondents ($f = 100$) as $\frac{150}{100}$. By canceling common factors in the numerator and denominator, it is possible to reduce a ratio to its simplest form, for example, $\frac{150}{100} = \frac{3}{2}$. (There were 3 black respondents for every 2 white respondents.)

The researcher might increase the clarity of his ratio by giving the base (the denominator) in some more understandable form. For instance, the *sex ratio* often employed by demographers, who seek to compare the number of males and females in any given population, is generally given as the number of males per 100 females.

To illustrate, if the ratio of males to females is $\frac{150}{50}$, there would be 150 males for 50 females (or reducing, 3 males for every 1 female). To obtain the conventional version of the sex ratio, we would multiply the above ratio by 100. Then,

$$\text{Sex ratio} = (100)\frac{f\,\text{males}}{f\,\text{females}} = \frac{(100)150}{50} = 300$$

It turns out then that there were 300 males in the population for every 100 females.

Ratios are no longer extensively used in social research, perhaps for the following reasons:

1. A large number of ratios are needed to describe distributions having many categories of analysis.
2. It may be difficult to compare ratios based on very large numbers.
3. Some social researchers prefer to avoid the fractions or decimals generated by ratios.

Rates

Another kind of ratio—one which tends to be more widely used by social researchers—is known as a *rate*. Sociologists often analyze populations regarding rates of reproduction, death, crime, divorce, marriage, and the like. However, while most other ratios compare the number of cases in any subgroup (category) with the number of cases in any other subgroup (category), rates indicate comparisons between the number of *actual* cases and the number of *potential* cases. For instance, to determine the birth rate for a given population, we might show the number of *actual* live births among females of childbearing age (those members of the population who are exposed to the risk of childbearing and therefore represent potential cases). Similarly, to ascertain the divorce rate, we might compare the number of actual divorces against the number of marriages which occur during some period of time (for example, one year). Rates are often given in terms of a base having 1000 potential cases. Thus, birth rates are given as the number of births per 1000 females; divorce rates might be expressed in terms of the number of divorces per 1000 marriages. Then, if 500 births occur among 4000 women of childbearing age,

$$\text{Birth rate} = (1000) \frac{f \text{ actual cases}}{f \text{ potential cases}} = \frac{(1000)500}{4000} = 125$$

It turns out there were 125 live births per every 1000 women of childbearing age.

So far we have discussed rates that might be useful in making comparisons between different populations. For instance, we might seek to compare birth rates between blacks and whites, between middle-class and lower-class females, among religious groups or entire societies, and so on. Another kind of rate, *rate of change,* can be used to compare the same population at two points in time. In computing rate of change, we compare the actual change between time 1 and time 2 with the size of time period 1 serving as a base. Thus, a population which increases from 20,000 to 30,000 between 1960 and 1970 would experience a rate of change:

$$\frac{(100) \text{ time } 2f - \text{time } 1f}{\text{time } 1f} = \frac{(100)30,000 - 20,000}{20,000} = 50\%$$

TABLE 2.4
The distribution of religious preferences shown three ways

Religion	f	Religion	f	Religion	f
Protestant	30	Catholic	20	Jewish	10
Catholic	20	Jewish	10	Protestant	30
Jewish	10	Protestant	30	Catholic	20
Total	60	Total	60	Total	60

In other words, there was a population increase of 50 percent over the period 1960 to 1970.

Notice that a rate of change can be *negative* to indicate a decrease in size over any given period. For instance, if a population changes from 15,000 to 5,000 over a period of time, the rate of change would be

$$\frac{(100)5,000 - 15,000}{15,000} = -67\%$$

SIMPLE FREQUENCY DISTRIBUTIONS OF ORDINAL AND INTERVAL DATA

Since nominal data are labeled rather than graded or scaled, the categories of nominal-level distributions do not have to be listed in any particular order. Thus, the data of religious preferences shown in Table 2.4 are presented in three different, yet equally acceptable arrangements.

In contrast, the categories or score values in ordinal or interval distributions represent the degree to which a particular characteristic is present. The listing of such categories or score values in simple frequency distributions must be made to reflect that order.

For this reason, ordinal and interval categories are always arranged in order from their highest to lowest values. For instance, we might list the categories of social class from upper to lower (upper, middle, lower), or we might post the results of a biology midterm examination in consecutive order from the highest grade to the lowest grade.

Disturbing the order of ordinal and interval categories reduces the readability of the researcher's findings. This effect can be seen in

TABLE 2.5
A frequency distribution of attitudes toward nuclear energy on a college campus: correct and incorrect presentations

Attitude Toward Nuclear Energy	f	Attitude Toward Nuclear Energy	f
Slightly favorable	2	Strongly favorable	0
Somewhat unfavorable	10	Somewhat favorable	1
Strongly favorable	0	Slightly favorable	2
Slightly unfavorable	4	Slightly unfavorable	4
Strongly unfavorable	21	Somewhat unfavorable	10
Somewhat favorable	1	Strongly unfavorable	21
Total	38	Total	38
INCORRECT		CORRECT	

Table 2.5, where both the "correct" and "incorrect" versions of a distribution of "Attitudes Toward Nuclear Energy on a College Campus" have been presented. Which version do you find easier to read?

GROUPED FREQUENCY DISTRIBUTIONS OF INTERVAL DATA

Interval-level scores are sometimes spread over a wide range (highest minus lowest score), making the resultant simple frequency distribution long and hard to read. When such instances occur, few cases may fall at each score value, and the group pattern becomes blurred. To illustrate, the distribution set up in Table 2.6 contains values varying from 50 to 99 and runs almost four columns in length.

In order to clarify our presentation, we might construct a *grouped frequency distribution* by condensing the separate scores into a number of smaller categories or groups, each containing more than one score value. Each category or group in a grouped distribution is known as a *class interval,* whose *size* is determined by the number of score values it contains.

The examination grades for 71 students originally presented in Table 2.6 are rearranged in a grouped frequency distribution, shown in Table 2.7. Here we find 10 class intervals, each having size 5. Thus, the highest class interval (95–99) contains the five score values 95, 96, 97, 98, and 99. Similarly, the interval 70–74 is of size 5 and contains the score values 70, 71, 72, 73, and 74.

Class Limits

In accordance with its size, each class interval has an *upper limit* and a *lower limit.* At first glance, the highest and lowest scores in any given category *seem* to be these limits. Thus, we might reasonably expect the upper and lower limits of the interval 60–64 to be 64 and 60, respectively. In this case, however, we would be *wrong,* since 64 and 60 are actually not the limits of interval 60–64.

TABLE 2.6
Frequency distribution of final examination grades for 71 students

Grade	f	Grade	f	Grade	f	Grade	f
99	0	85	2	71	4	57	0
98	1	84	1	70	9	56	1
97	0	83	0	69	3	55	0
96	1	82	3	68	5	54	1
95	1	81	1	67	1	53	0
94	0	80	2	66	3	52	1
93	0	79	8	65	0	51	1
92	1	78	1	64	1	50	1
91	1	77	0	63	2	Total	71
90	0	76	2	62	0		
89	1	75	1	61	0		
88	0	74	1	60	2		
87	1	73	1	59	3		
86	0	72	2	58	1		

TABLE 2.7
Grouped frequency distribution of final examination grades for 71 students

Class Interval	f
95–99	3
90–94	2
85–89	4
80–84	7
75–79	12
70–74	17
65–69	12
60–64	5
55–59	5
50–54	4
Total	71

Many readers are asking, "Why not?" To provide an answer, let us examine a problem that might arise if we were to define class limits in terms of the highest and lowest scores in any interval. Suppose we tried placing numbers containing fractional values (decimal fractions) in the frequency distribution set up in Table 2.7. Where would we categorize the score 62.3? Most of us would agree that it belongs in the interval 60–64. But what about the score 69.4? How about the number 54.2 or 94.6? The reader might notice that the highest and lowest scores in an interval will leave gaps between adjacent groups, so that some fractional values cannot be assigned to any class interval in the distribution and must be excluded altogether.

Unlike the highest and lowest scores in an interval, *class limits* are located at the point halfway between adjacent class intervals and, therefore, serve to close the gap between them (see Figure 2.1). Thus, the upper limit of the interval 90–94 is 94.5, and the lower limit of the interval 95–99 is also 94.5. Likewise, 59.5 serves as the upper limit of the interval 55–59 and as the lower limit of the interval 60–64. The reader might ask: what about the value

FIGURE 2.1
The highest and lowest scores versus *the upper and lower limits of the class interval 90–94*

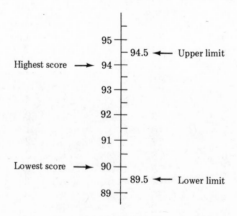

59.5—the value that falls *exactly* halfway between neighboring class intervals? Should we include this score in the interval 55–59 or in the interval 60–64? This problem is generally handled by *rounding to the closest even number*. For instance, 59.5 would be placed in the interval 60–64; 84.5 would be included in the interval 80–84. As we shall see, the position of class limits must be determined in order to work through certain statistical procedures.

The Midpoint

Another characteristic of any class interval is its *midpoint,* which we define as the middlemost score value in the class interval. A quick and simple method to find a midpoint is to look for the point at which any given interval can be divided into two equal parts. Taking some illustrations, 50 is the midpoint of the interval 48–52; 3.5 is the midpoint of the interval 2–5. The midpoint can also be computed from the lowest and highest scores in any interval. To illustrate, the midpoint of the interval 48–52 is

$$\frac{\text{lowest score} + \text{highest score}}{2} = \frac{48 + 52}{2} = 50$$

Determining the Number of Intervals

To present interval data in a grouped frequency distribution, the social researcher must consider the number of categories he wishes to employ. Texts generally advise using as few as 5 intervals to as many as 20 intervals. In this regard, it would be wise to remember that grouped frequency distributions are employed to reveal or emphasize a group pattern. Either too many or too few class intervals may blur that pattern and thereby work against the researcher who seeks to add clarity to his analysis. In addition, reducing the individual score values to an unnecessarily small number of intervals may sacrifice too much precision—precision which was originally attained by knowing the identity of individual scores in the distribution. In sum, then, the researcher generally makes his decision as to the number of intervals based on his own set of data and his personal objectives, factors which may vary considerably from one research situation to another.

CUMULATIVE DISTRIBUTIONS

It is sometimes desirable to present frequencies in a cumulative fashion, especially when we seek to locate the position of one case relative to overall group performance. *Cumulative frequencies* are defined as the total number of cases having any given score *or a score that is lower*. Thus, the cumulative frequency *cf* for any category (or class interval) is obtained by adding the frequency in that category to the total frequency for all categories below it. In the case of the college board scores in Table 2.8, we see that the frequency *f* associated with the class interval 301–350 is 12. This is also the cu-

Description

TABLE 2.8
*Cumulative frequency
distribution of college board
scores for 336 students*

Class Interval		f	cf
751–800		6	336
701–750		25	330
651–700		31	305
601–650		30	274
551–600		35	244
501–550		55	209
451–500		61	154
401–450		48	93
351–400		33	45
301–350		12	12
	Total	336	

mulative frequency for this interval since no member of the group scored below 301. The frequency in the next class interval 351–400 is 33, while the cumulative frequency for this interval is 45 (33 + 12). Therefore, we learn that 33 students earned college board scores between 351 and 400, but that 45 students received scores of 400 *or lower*. We might continue this procedure, obtaining cumulative frequencies for all class intervals, until we arrive at the topmost entry 751–800, whose cumulative frequency (336) is equal to the total number of cases since no member of the group scored above 800.

In addition to cumulative frequency, we can also construct a distribution which indicates *cumulative percentage* (c%), the percent of cases having any score or a score that is lower. To calculate cumulative percentage, we modify the formula for percentage (%) introduced earlier in this chapter as follows:

$$c\% = (100)\frac{cf}{N}$$

where

cf = the cumulative frequency in any category
N = the total number of cases in the distribution

Applying the foregoing formula to the data in Table 2.8, we find that the percent of students who scored 350 or lower was

$$c\% = (100)\frac{12}{336}$$

$$= (100).0357$$
$$= 3.57$$

The percent who scored 400 or lower was

$$c\% = (100)\frac{45}{336}$$

$$= (100).1339$$
$$= 13.39$$

The percent who scored 450 or lower was

$$c\% = (100)\,\frac{93}{336}$$
$$= (100).2768$$
$$= 27.68$$

A cumulative percentage distribution based on the data in Table 2.8 is shown in Table 2.9.

PERCENTILE RANK   Suppose you received a score of 80 on a statistics examination. In order to determine exactly how well you have done, it would probably be helpful to know how your score of 80 compares with the scores of others in your class who took the same exam. Did most of the other students score in the 80s and 90s? If so, your own grade may not be terribly high. Or, did most of the others receive scores in the 60s and 70s? If so, a score of 80 may very well fall among the highest in your class.

With the aid of the cumulative percentage distribution, we can make precise comparisons between any individual case and the group in which it occurs. Specifically, we can find the *percentile rank* of a score, a single number which indicates the percent of cases in a distribution which fall below any given score. For instance, if a score of 80 has a percentile rank of 95, then 95 percent of the students in this statistics class received exam scores lower than 80 (only 5 percent scored above 80). If, however, a score of 80 has a percentile rank of 45, then only 45 percent received exam scores below 80 (55 percent scored above it). By formula,

$$\text{Percentile rank} = \begin{array}{c} c\% \text{ below}\\ \text{the lower}\\ \text{limit of}\\ \text{critical}\\ \text{interval} \end{array} + \left[\frac{\text{score} - \begin{array}{c}\text{lower limit of}\\ \text{critical interval}\end{array}}{\text{size of critical interval}}\left(\begin{array}{c}\% \text{ in}\\ \text{critical}\\ \text{interval}\end{array}\right)\right]$$

To illustrate the procedure for obtaining percentile rank, let us find the percentile rank for a score of 620 in the distribution in

TABLE 2.9
Cumulative percentage distribution of college board scores for 336 students (based on data in Table 2.8)

Class Interval	cf	c%
751–800	336	100%
701–750	330	98.21
651–700	305	90.77
601–650	274	81.55
551–600	244	72.62
501–550	209	62.20
451–500	154	45.83
401–450	93	27.68
351–400	45	13.39
301–350	12	3.57

Table 2.8. Before applying the formula, we must first locate the *critical interval,* that class interval in which a score of 620 appears. As shown below, the critical interval for the present problem is 601–650:

Class Interval

751–800
701–750
651–700
601–650   ← Class interval in which
551–600   a score of 620 occurs
501–550
451–500
401–450
351–400
301–350

There are several characteristics of the critical interval which we must determine before applying the formula for percentile rank:

1. The lower limit of the critical interval. This is the point that lies midway between the critical interval, 601–650, and the class interval immediately below it, 551–600. The lower limit of 601–650 is 600.5.
2. The size of the critical interval. This is determined by the number of score values within the class interval, 601–650. The size of the critical interval is 50, since it contains 50 score values from 601 to 650.
3. The percent within the critical interval. To determine the percent within any class interval, we divide the number of cases in that class interval (f) by the total number of cases in the distribution N, and multiply our answer by 100. By formula,

$$\% = (100)\frac{f}{N}$$

$$= (100)\frac{30}{336}$$

$$= (100).089$$
$$= 8.93$$

Therefore, we see that 8.93 percent of these college board scores fell within the class interval, 601–650.

4. The cumulative percentage below the lower limit of the critical interval. We can read $c\%$ directly from the cumulative percentage distribution in Table 2.9. Moving up the $c\%$ column of the table, we see that 72.62 percent of the scores fall *below* the critical interval. This is the cumulative percent-

age associated with the class interval that falls immediately below the critical interval.

We are now prepared to apply the formula for percentile rank:

$$\text{Percentile rank} = 72.62 + \left[\frac{620 - 600.5}{50} (8.93) \right]$$

$$= 72.62 + \left[\frac{19.50}{50} (8.93) \right]$$

$$= 72.62 + (.39)(8.93)$$

$$= 72.62 + 3.48$$

$$= 76.10$$

It turns out that slightly more than 76 percent received a score lower than 620. Only 23.90 percent scored higher.

As another illustration, let us find the percentile rank for a score of 92 in the following distribution of scores:

Class Interval	f	cf	c%
90–99	6	49	100%
80–89	8	43	87.76
70–79	12	35	71.43
60–69	10	23	46.94
50–59	7	13	26.53
40–49	6	6	12.24
	N = 49		

As shown below, the critical interval for a score of 92 is 90–99:

Class Interval	
90–99	← Class interval in which
80–89	a score of 92 occurs
70–79	
60–69	
50–59	
40–49	

The following are the characteristics of the critical interval which we must determine:

1. The lower limit of the critical interval is 89.5.
2. The size of the critical interval is 10, since there are 10 score values within it from 90 to 99 (90 91 92 93 94 95 96 97 98 99).
3. The percent within the critical interval is 12.24. By formula,

$$\% = (100) \frac{f}{N}$$

$$= (100) \frac{6}{49}$$

$$= (100).1224$$

$$= 12.24$$

4. The cumulative percentage below the lower limit of the critical interval can be found from the c% column by referring to the class interval immediately below the critical interval. The cumulative percentage associated with the class interval 80–89 is 87.76.

We are now ready to substitute in the formula for percentile rank:

$$\text{Percentile rank} = 87.76 + \left[\frac{92 - 89.5}{10} (12.24) \right]$$

$$= 87.76 + \left[\frac{2.50}{10} (12.24) \right]$$

$$= 87.76 + (.25)(12.24)$$

$$= 87.76 + 3.06$$

$$= 90.82$$

Almost 91 percent received a score lower than 92. Only 9.18 percent got a score that was higher.

The scale of percentile ranks consists of 100 units. There are certain percentile ranks along this scale which have specific names. *Deciles* divide the scale of percentile ranks by tens. Thus, if a score is located at the first decile (percentile rank = 10), we know that 10 percent of the cases fall below it; if a score is at the second decile (percentile rank = 20), then 20 percent of the cases fall below it, and so on. Percentile ranks which divide the scale by twenty-fives are known as *quartiles*. If a score is located at the first quartile (percentile rank = 25), we know that 25 percent of the cases fall below it; if

Figure 2.2
Scale of percentile ranks divided by deciles and quartiles

Percentile rank	Decile	Quartile
90 =	9th	
85		
80 =	8th	
75 =		3rd
70 =	7th	
65		
60 =	6th	
55		
50 =	5th	2d
45		
40 =	4th	
35		
30 =	3rd	
25 =		1st
20 =	2nd	
15		
10 =	1st	

a score is at the second quartile (percentile rank = 50), 50 percent of the cases fall below it; and if a score is at the third quartile (percentile rank = 75), 75 percent of the cases fall below it (see Figure 2.2).

SUMMARY

In this chapter we were introduced to some of the basic techniques used by the social researcher to organize the jumble of raw numbers that he or she collects from respondents. Frequency distributions and methods for comparing such distributions of nominal data (proportions, percentages, ratios, and rates) were discussed and illustrated. With respect to ordinal and interval data, the characteristics of simple, grouped, and cumulative frequency distributions were examined. Finally, the procedure for obtaining the percentile rank of a raw score was presented.

PROBLEMS

1. From the following table representing achievement for television viewers and nonviewers, find (a) the percent of nonviewers who are high achievers, (b) the percent of viewers who are high achievers, (c) the proportion of nonviewers who are high achievers, and (d) the proportion of viewers who are high achievers.

Achievement for Television Viewers and Nonviewers

	Viewing Status	
	Nonviewers	*Viewers*
Achievement	*f*	*f*
High achievers	93	46
Low achievers	90	127
Total	183	173

2. From the following table representing family structure for black and white children, find (a) the percent of black children

Family Structure for Black and White Children

	Race of Child	
	Black	*White*
Family Structure	*f*	*f*
One parent	53	59
Two parents	130	167
Total	183	226

having two-parent families, (b) the percent of white children having two-parent families, (c) the proportion of black children having two-parent families, and (d) the proportion of white children having two-parent families.

3. In a group of 4 high achievers and 24 low achievers, what is the ratio of high-to-low achievers?

4. In a group of 125 males and 80 females, what is the ratio of males to females?

5. In a group of 15 black children and 20 white children, what is the ratio of blacks to whites?

6. If 300 live births occur among 3500 women of childbearing age, what is the birthrate?

7. What is the rate of change for population increase from 15,000 in 1950 to 25,000 in 1970?

8. Convert the following distribution of scores into a grouped frequency distribution containing four class intervals, and (a) determine the size of class intervals, (b) indicate the upper and lower limits of each class interval, (c) identify the midpoint of each class interval, (d) find the cumulative frequency for each class interval, and (e) find the cumulative percentage for each class interval.

Score Value	f
12	3
11	4
10	4
9	5
8	6
7	5
6	4
5	3
4	2
3	1
2	1
1	2
	$N = 40$

9. Convert the following distribution of scores into a grouped frequency distribution containing five class intervals, and (a) determine the size of class intervals, (b) indicate the upper and lower limits of each class interval, (c) identify the midpoint of each class interval, (d) find the cumulative frequency for each class interval, and (e) find the cumulative percentage for each class interval.

Score Value	f
100	4
99	6
98	7
97	9
96	12
95	10
94	6
93	5
92	3
91	2
	N = 64

10. In the following distribution of scores, find the percentile rank
for (a) a score of 75, and (b) a score of 52.

Class Interval	f	cf
90–99	6	48
80–89	9	42
70–79	10	33
60–69	10	23
50–59	8	13
40–49	5	5
	N = 48	

11. In the following distribution of scores, find the percentile rank
for (a) a score of 36, and (b) a score of 18.

Class Interval	f
40–44	5
35–39	5
30–34	8
25–29	9
20–24	10
15–19	8
10–14	6
5–9	5
	N = 56

TERMS TO
REMEMBER

Frequency distribution
Proportion
Percentage
Ratio
Rate
Grouped frequency distribution
Class interval

Class limits
Midpoint
Cumulative frequency
Cumulative percentage
Percentile rank
Deciles
Quartiles

3 Graphic Presentations

COLUMNS OF NUMBERS have been known to evoke fear, boredom, apathy, and misunderstanding. While some persons seem to "tune out" statistical information presented in tabular form, they may pay close attention to the same data presented in graphic or picture form. As a result, many commercial researchers and popular authors prefer to use graphs as opposed to tables. For similar reasons, social researchers often use graphs such as pie charts, bar graphs, and frequency polygons in an effort to increase the readability of their findings.

PIE CHARTS

One of the simplest graphic methods can be found in the *pie chart,* a circular graph whose pieces add up to 100 percent. Pie charts are particularly useful for visualizing differences in frequencies among a few nominal-level categories. To illustrate, Figure 3.1 presents the urban, suburban, or rural backgrounds of a population of 2000 college students. Notice that 70 percent of these students come from suburban areas, whereas only 12 percent come from rural areas.

BAR GRAPHS

The pie chart provides a quick and easy illustration of data which can be divided into a few categories. By comparison, the *bar graph* (or *histogram*) can accommodate any number of categories at any level of measurement and, therefore, is more widely used in social research.

　　Let us examine the bar graph in Figure 3.2 which illustrates a frequency distribution of social classes. This bar graph is set up fol-

36

FIGURE 3.1
Urban, suburban, and rural backgrounds of a population of 2000 college students

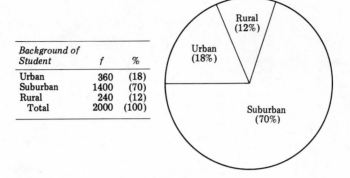

Background of Student	f	%
Urban	360	(18)
Suburban	1400	(70)
Rural	240	(12)
Total	2000	(100)

lowing the standard arrangement: a horizontal base line (or *x* axis) along which the score values or categories (in this example, social classes) are marked off and a vertical line (*y* axis) along the side of the figure representing the frequencies for each score value or category. (In the case of grouped data, the midpoints of the class intervals are arranged along the horizontal base line.) Notice that rectangular bars give the frequencies for the range of score values. The taller the bar, the greater the frequency of occurrence.

In Figure 3.2, the rectangular bars of the graph have been joined in order to emphasize differing degrees of social status as represented by social class differences. In addition, social classes have been plotted on the base line in *ascending* order from lower-lower to upper-upper. This is the conventional arrangement for constructing ordinal- and interval-level bar graphs.

In drawing a bar graph of nominal data, however, the bars should be *separated* rather than joined to avoid implying continuity

FIGURE 3.2
Bar graph of a distribution of social classes

Social class	f
Upper-upper	5
Lower-upper	14
Upper-middle	23
Lower-middle	45
Upper-lower	38
Lower-lower	25
Total	150

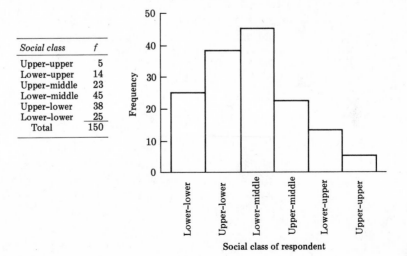

FIGURE 3.3
*Bar graph of an occupa-
tional distribution*

Occupation	f
Craftsmen	52
Unskilled labor	65
Managerial	29
Clerical	34
Total	180

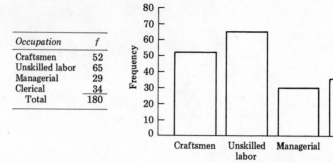

Occupation of respondent

among the categories. Moreover, nominal-level categories can be arranged in any order along the horizontal base line. Figure 3.3 illustrates such characteristics of nominal-level bar graphs.

FREQUENCY POLYGONS

Another commonly employed graphic method is the *frequency polygon*. Although the frequency polygon can accommodate a wide variety of categories, it tends to stress *continuity* along a scale rather
than *differentness* and, therefore, is particularly useful for depicting
ordinal and interval data. This is because frequencies are indicated
by a series of points placed over the score values or midpoints of each
class interval. Adjacent points are connected with a straight line,
which is dropped to the base line at either end. As shown in Figure
3.4, the height of each point or dot indicates frequency of occurrence.

In order to graph cumulative frequencies (or cumulative percentages), it is possible to construct a *cumulative frequency polygon*.
As shown in Figure 3.5, cumulative frequencies are arranged along
the vertical line of the graph and are indicated by the height of
points above the horizontal base line. Unlike a regular frequency
polygon, however, the straight line connecting all points in the cumulative frequency polygon cannot be dropped back to the base line,
since the cumulative frequencies being represented are a product of

FIGURE 3.4
*Frequency polygon of a dis-
tribution of IQ scores*

Class interval	f
136–145	11
126–135	16
116–125	29
106–115	40
96–105	44
86–95	25
76–85	13
Total	178

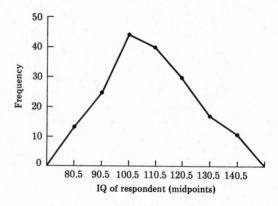

IQ of respondent (midpoints)

FIGURE 3.5
Cumulative frequency polygon for the data in Table 2.8

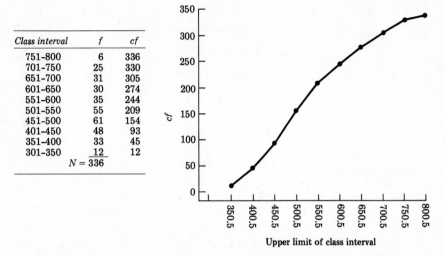

Class interval	f	cf
751–800	6	336
701–750	25	330
651–700	31	305
601–650	30	274
551–600	35	244
501–550	55	209
451–500	61	154
401–450	48	93
351–400	33	45
301–350	12	12
	N = 336	

Upper limit of class interval

successive additions. Any given cumulative frequency is never less (and is usually more) than the preceding cumulative frequency. Also unlike a regular frequency polygon, the points in a cumulative graph are plotted above the upper limits of class intervals rather than at their midpoints. This is because cumulative frequency represents the total number of cases *both within and below* a particular class interval.

CONSTRUCTING BAR GRAPHS AND FREQUENCY POLYGONS

The following rules and procedures can be applied to the construction of both bar diagrams and frequency polygons:

1. As a matter of tradition and to avoid confusion, the researcher always arranges score values along the horizontal base line and the frequencies (or percent of cases) along the vertical line.
2. Every graph is completely labeled. The horizontal base line should be labeled as to characteristic (for example, age of respondent); the vertical line should be labeled according to what is being represented (either "frequencies" or "percents") and the numerical values of points along the scale. In addition, the graph should be given a title, which indicates the nature of the data being illustrated.
3. In constructing a graph, the length of the vertical line should be about 75 percent of the length of the horizontal base line. This arrangement provides a relatively standard way to draw graphs and minimizes a source of potential confusion.
4. The first point on the vertical line—that point at which the vertical line meets the horizontal line—should always begin with zero, since any other arrangement might give a distorted picture of the data.

Description

FIGURE 3.6
Some variations in kurtosis among symmetrical distributions

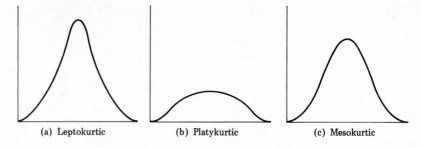

(a) Leptokurtic   (b) Platykurtic   (c) Mesokurtic

THE SHAPE OF A FREQUENCY DISTRIBUTION

Graphic methods can help us to visualize the variety of shapes and forms taken by frequency distributions. Some distributions are *symmetrical*—folding the curve at the center creates two identical halves. Therefore, such distributions contain the same number of extreme score values in both directions, high and low. Other distributions are said to be *skewed* and have more extreme cases in one direction than the other.

There is considerable variation among symmetrical distributions. For instance, they can differ markedly in terms of *peakedness* (or *kurtosis*). Some symmetrical distributions, as in Figure 3.6 (a), are quite peaked or tall (called *leptokurtic*); others, as in Figure 3.6 (b), are rather flat (called *platykurtic*); still others are neither very peaked nor very flat (called *mesokurtic*). One kind of mesokurtic symmetrical distribution as shown in Figure 3.6 (c), the *normal curve,* has special significance for social research and will be discussed in some detail in Chapter 6.

There is a variety of skewed or asymmetrical distributions. When skewness exists so that scores pile up in one direction, the distribution will have a pronounced "tail." The position of this tail indicates where the relatively few extreme scores are located and determines the *direction* of skewness.

Distribution (a) in Figure 3.7 is *negatively skewed* (skewed to the left), since it has a much longer tail on the left than the right. This distribution shows that most respondents received high scores, but only a few obtained low scores. If this were the distribution of grades on a final examination, we could say that most students did quite well, and a few did poorly.

Next look at distribution (b), whose tail is situated to the right.

FIGURE 3.7
Three distributions representing direction of skewness

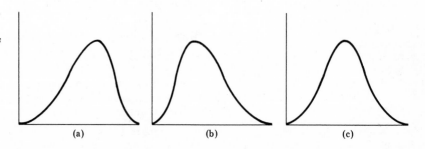

(a)   (b)   (c)

Since skewness is indicated by the direction of the tail, we can say that the distribution is *positively skewed* (skewed to the right). The final examination grades for the students in our hypothetical classroom would be quite low!

Finally, let us examine distribution (c), which contains two identical tails. In such a case, there is the same number of extreme scores in both directions. The distribution is not at all skewed, but is perfectly symmetrical. If this were the distribution of grades on our final examination, we would have a large number of more or less average students and few students receiving high and low grades.

SUMMARY

Graphic presentations of data can be used to increase the readability of research findings. Our discussion of graphic presentations included pie charts, bar graphs, and frequency polygons. Pie charts provide a simple illustration of data, which are divisable into a few categories. Bar graphs are more widely used, since they can accommodate any number of categories. Frequency polygons also accommodate a wide range of categories, but are especially useful for ordinal and interval data since they stress continuity along a scale.

Variations in the shape of distributions can be characterized in terms of symmetry or, if containing more extreme cases in one direction than the other, in terms of positive or negative skewness.

PROBLEMS

1. Use pie charts to depict the following information about the students at a hypothetical college:

Religious Preferences of Students	f	%
Protestant	80	(40)
Catholic	80	(40)
Jewish	40	(20)
Total	200	(100)

Sex	f	%
Male	100	(50)
Female	100	(50)
Total	200	(100)

Citizenship of Students	f	%
U.S.A.	180	(90)
Other	20	(10)
Total	200	(100)

2. Depict the following data in a bar graph:

Country of Origin of Foreign Students	f
Iran	5
China	7
England	2
Egypt	5
Greece	3
Other	4

3. On graph paper, draw both a bar graph and a frequency polygon to illustrate the following distribution of IQ scores:

Class Interval	f
103–104	2
101–102	4
99–100	22
97–98	10
95–96	6
	N = 44

4. On graph paper, draw a cumulative frequency polygon to represent the following grades on a final exam:

Class Interval	f	cf
91–100	3	27
81–90	6	24
71–80	12	18
61–70	4	6
51–60	2	2
	N = 27	

TERMS TO REMEMBER

Pie chart
Bar graph
Frequency polygon
Cumulative frequency polygon
Kurtosis
   Leptokurtic
   Platykurtic
   Mesokurtic
Positive and negative skewness

4 • Measures of Central Tendency

RESEARCHERS IN MANY fields have used the term "average" to ask such questions as: what is the *average* income earned by high school and college graduates? How many cigarettes are smoked by the *average* teenager? What is the grade-point *average* of the female college student? On the *average,* how many automobile accidents happen as the direct result of alcohol or drugs?

A useful way to describe a group as a whole is to find a single number that represents what is "average" or "typical" of that set of data. In social research, such a value is known as a *measure of central tendency,* since it is generally located toward the middle or center of a distribution where most of the data tend to be concentrated.

What the layman means by the term "average" is often vague and even confusing. The social researcher's conception is much more precise than the popular usage; it is expressed numerically as one of several different kinds of measures of "average" or central tendency which may take on quite different numerical values in the same set of data. Only the three best known measures of central tendency are discussed here: the *mode,* the *median,* and the *mean.*

THE MODE

To obtain the mode (Mo) simply find the score or category that occurs most often in a distribution. The mode can be easily found by inspection, rather than by computation. For instance, in the set of scores ①, 2, 3, ①, ①, 6, 5, 4, ①, 4, 4, 3, the mode is 1, since it is the number that occurs more than any other score in the set (it occurs four times).

In the case of a simple frequency distribution where the score values and frequencies are presented in separate columns, the mode is the score value that appears most often in the frequency column of

Score Value	f
7	2
6	3
5	4
Mo → 4	5
3	4
2	3
1	2
Total	23

the table. Therefore, Mo = 4 in the simple frequency distribution located in Table 4.1.

Some frequency distributions contain two or more modes. In the following set of data, for example, the scores 2 and 6 *both* occur most often: 6,6,7,2,6,1,2,3,2,4. Graphically, such distributions have two points of maximum frequency, suggestive of the humps on a camel's back. These distributions are referred to as being *bimodal* in contrast to the more common *unimodal* variety, which has only a single hump or point of maximum frequency (see Figure 4.1).

THE MEDIAN

When ordinal or interval data are arranged in order of size, it becomes possible to locate the median (Mdn), the *middlemost* point in a distribution. Therefore, the median is regarded as the measure of central tendency that cuts the distribution into two equal parts.

If we have an odd number of cases, then the median will be the case that falls exactly in the middle of the distribution. The position of the median value can be located by inspection or by the formula,

$$\text{Position of median} = \frac{N + 1}{2}$$

Thus, 16 is the median value for the scores 11, 12, 13, ⑯, 17, 20, 25; this is the case that divides the numbers, so that there are three scores on either side of it. According to the formula $(7 + 1)/2$, we see that the median 16 is the fourth score in the distribution counting from either end.

If the number of cases is even, the median is always that *point* above which 50 percent of the cases fall and below which 50 percent of the cases fall. For an even number of cases, there will be two

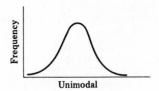

Unimodal

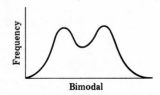

Bimodal

middle cases. To illustrate, the numbers 16 and 17 represent the middle cases for the following data: 11, 12, 13, ⑯, ⑰, 20, 25, 26. By the formula $(8 + 1)/2 = 4.5$, the median will fall midway between the fourth and fifth cases; the middlemost point in this distribution turns out to be 16.5, since it lies halfway between 16 and 17, the fourth and fifth scores in the set. Likewise, the median is 9 in the data 2, 5, ⑧, ⑩, 11, 12, again because it is located exactly midway between the two middle cases $(6 + 1)/2 = 3.5$.

Another circumstance must be explained and illustrated—we may be asked to find the median from data containing several middle scores having identical numerical values. The solution is simple—that numerical value becomes the median. Therefore, in the data 11, 12, 13, ⑯, ⑯, ⑯, 25, 26, 27, the median case is 16, although it occurs more than once.

Obtaining the Median from a Simple Frequency Distribution

To find the median for data arranged in the form of a simple frequency distribution, we start with the procedure just presented. In the case of Table 4.1,

$$\text{Position of median} = \frac{23 + 1}{2}$$

$$= \frac{24}{2}$$

$$= 12$$

The median turns out to be the twelfth score in this frequency distribution. To help locate the twelfth score, we might construct a cumulative frequency distribution as shown in the third column of Table 4.2 (for a small number of scores, this can be done in your head). Beginning with the lowest score value, we add frequencies until coming to the twelfth score in the distribution. In the present example, the median score value is 4.

THE MEAN

By far the most commonly used measure of central tendency, the arithmetic mean $\overline{X}$, can be obtained by adding up a set of scores and

TABLE 4.2
Finding the median for a simple frequency distribution

Score Value	f	cf
7	2	23
6	3	21
5	4	18
Mdn → 4	5	14
3	4	9
2	3	5
1	2	2
Total	23	

TABLE 4.3
Calculating the mean: an illustration

Respondent	X(IQ)	
Bonnie	125	
Steve	92	$\overline{X} = \dfrac{\Sigma X}{N}$
Jamie	72	
Michael	126	
Joan	120	$= \dfrac{864}{8}$
Jim	99	
Andrea	130	
Megan	100	$= 108$
	$\Sigma X = 864$	

dividing by the number of scores. Therefore, we define the mean more formally as *the sum of a set of scores divided by the total number of scores in the set.* By formula,

$$\overline{X} = \frac{\Sigma X}{N}$$

where

$\overline{X}$ = the mean (read as X bar)
Σ = the sum (expressed as the Greek capital letter sigma)[1]
X = a raw score in a set of scores
N = the total number of scores in a set

Using the formula above, we learn that the mean IQ for the eight respondents listed in Table 4.3 is 108.

Unlike the mode, the mean is not always the score that occurs most often. Unlike the median, it is not necessarily the middlemost point in a distribution. Then, what does the *mean* mean? How can it be interpreted? As we shall see, the mean can be regarded as the "center of gravity," the point in any distribution around which the positive and negative deviations balance. In order to understand this characteristic of the mean, we must first understand the concept of *deviation* which indicates the distance and direction of of any raw score from the mean. To find a deviation score we simply subtract the mean from any given raw score. According to the formula,

$$x = X - \overline{X}$$

where

x = the deviation score (always symbolized by small x)
X = any raw score in the distribution
$\overline{X}$ = the mean

[1] The Greek capital letter sigma (Σ) will be encountered many times throughout the text. It simply indicates that we must *sum* or add up what follows. In the present example, ΣX indicates adding up the raw scores.

TABLE 4.4
Deviations of a set of raw scores from X̄

X	x	
9	+3⎫	+5
8	+2⎭	
6	0	$\overline{X} = 6$
4	−2⎫	−5
3	−3⎭	

Since $\overline{X} = 6$ for the set of raw scores 9, 8, 6, 4, and 3, the raw score 9 lies exactly three raw score units *above* the mean of 6 (or $X - \overline{X} = 9 - 6 = +3$). Similarly, the raw score 4 lies two raw score units *below* the mean (or $X - \overline{X} = 4 - 6 = -2$). Conclusion: the greater the deviation x, the greater is the distance of that raw score from the mean of a distribution.

Considering the mean as a point of balance in the distribution, we can now say that the sum of the deviations that fall above the mean is equal in absolute value (ignoring the minus signs) to the sum of the deviations that fall below the mean. Let us return to an earlier example, the set of scores 9, 8, 6, 4, 3 in which $\overline{X} = 6$. If the mean for this distribution is the "center of gravity," then disregarding minus signs and adding together the positive deviations (deviations of raw scores 8 and 9) should equal adding together the negative deviations (deviations of raw scores 4 and 3). As shown in Table 4.4, this turns out to be the case since the sum of deviations below $\overline{X}$ (− 5) equals the sum of deviations above $\overline{X}$ (+ 5).

Taking another example, 4 is the mean for the numbers 1, 2, 3, 5, 6, and 7. We see that the sum of deviations below this score is − 6, while the sum of deviations above it is + 6. We shall return to the concept of deviation in Chapters 5 and 6.

Obtaining the Mean from a Simple Frequency Distribution

To obtain the mean for a small number of scores, the basic formula $\overline{X} = \Sigma X/N$ serves well. When we have a large number of cases, however, it may be more practical and less time-consuming to compute the mean from a frequency distribution by the formula

$$\overline{X} = \frac{\Sigma fX}{N}$$

where

$\overline{X}$ = the mean
X = a raw score value in the distribution
fX = a score multiplied by its frequency of occurrence
ΣfX = the sum of the fX's
N = the total number of scores

Table 4.5 illustrates the computation of the mean from a simple frequency distribution.

TABLE 4.5
Obtaining X̄ from a simple frequency distribution

X	f	fX	
8	2	16	
7	3	21	
6	5	30	
5	6	30	$\overline{X} = \dfrac{\Sigma fX}{N} = \dfrac{132}{28} = 4.71$
4	4	16	
3	4	12	
2	3	6	
1	1	1	
	N = 28	ΣfX = 132	

COMPARING THE MODE, MEDIAN, AND MEAN

The time comes when the social researcher chooses a measure of central tendency for a particular research situation. Will he or she employ the mode, the median, or the mean? The decision involves several factors including:

1. the level of measurement,
2. the shape or form of his distribution of data, and
3. the research objective.

Level of Measurement

Since the mode requires only a frequency count, it can be applied to any set of data at the nominal, ordinal, or interval level of measurement. For instance, we might determine that the modal category in a nominal-level measure of religious affiliation (Protestant, Catholic, or Jewish) is "Protestant," since the largest number of our respondents identify themselves as such. Similarly, we might learn that the largest number of students attending a particular university have a 2.5 grade-point average (Mo = 2.5).

The median requires an ordering of categories from highest to lowest. For this reason, it can only be obtained from ordinal or interval data, *not* from nominal data. To illustrate, we might find the median annual income is $17,000 among dentists in a small town. The result gives us a meaningful way to examine the central tendency in our data. By contrast, it would make little sense if we were to compute the median for scales of religious affiliation (Protestant, Catholic, or Jewish), sex (male or female), or country of origin (England, Poland, France, or Germany), when ranking or scaling has not been carried out.

The use of the mean is exclusively restricted to interval data. Applying it to ordinal or nominal data yields a meaningless result, generally not at all indicative of central tendency. What sense would it make to compute the mean for a distribution of religious affiliation or sex? Although less obvious, it is equally inappropriate to calculate a mean for data that can be ranked but not scored.

Shape of the Distribution

The shape or form of a distribution is another factor that can influence the researcher's choice of a measure of central tendency. In a

FIGURE 4.2
A unimodal, symmetrical distribution showing that the mode, median, and mean assume identical values

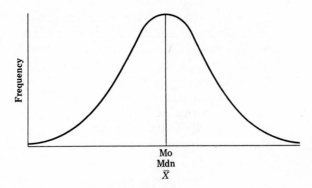

perfectly symmetrical, unimodal distribution, the mode, median, and mean will be identical, since the point of maximum frequency (Mo) is also the middlemost score (Mdn), as well as the "center of gravity" $(\overline{X})$. As shown in Figure 4.2, the measures of central tendency will coincide at the most central point, the "peak" of the symmetrical distribution.

When social researchers work with a symmetrical distribution, their choice of measure of central tendency is chiefly based on their particular research objectives and the level at which their data are measured. When they work with skewed distribution, however, this decision is very much influenced by the shape or form of their data.

As shown in Figure 4.3, the mode, median, and mean do not coincide in skewed distributions, *although their relative positions remain constant*—moving away from the "peak" and toward the "tail," the order is always from mode, to median, to mean. The mode falls closest to the "peak" of the curve, since this is the point where the most frequent scores occur. The mean, by contrast, is found closest to the "tail," where the relatively few extreme scores are located. For this reason, the mean score in the positively skewed distribution in Figure 4.3(a) lies toward the high score values; the mean in the negatively skewed distribution in Figure 4.3(b) falls close to the low score values.

FIGURE 4.3
The relative positions of measures of central tendency in (a) a positively skewed distribution and (b) a negatively skewed distribution

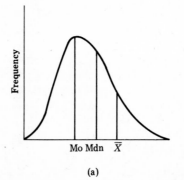

(a)

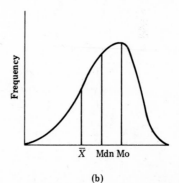

(b)

While the mean is very much influenced by extreme scores in either direction, the median is modified little, if at all, by changes in extreme values. This is because the mean considers all of the scores in any distribution, whereas the median (by definition) is concerned only with the numerical value of the score that falls at the middle-most position in a distribution. As illustrated below, changing an extreme score value from 10 in distribution A to 95 in distribution B does not at all modify the median value (Mdn = 7.5), whereas the mean shifts from 7.63 to 18.25:

distribution A; 5 6 6 7 8 9 10 10   Mdn = 7.5 $\overline{X}$ = 7.63
distribution B: 5 6 6 7 8 9 10 95   Mdn = 7.5 $\overline{X}$ = 18.25

In a skewed distribution, the median always falls somewhere between the mean and the mode. It is this characteristic that makes the median the most desirable measure of central tendency for describing a skewed distribution of scores. To illustrate this advantage of the median, let us turn to Table 4.6 and examine the "average" annual salary among employees working in a small corporation. If we were public relations practitioners hired by the corporation to give it a favorable public image, we would probably want to calculate the mean in order to show that the "average" employee makes $18,000 and is relatively well paid. On the other hand, if we were union representatives seeking to upgrade salary levels, we would probably want to employ the mode to demonstrate that the "average" salary is only $1000, an outrageously small amount. Finally, if we were social researchers seeking accurately to report the "average" salary among the employees in this corporation, we would wisely employ the median ($3000), since it falls between the other measures of central tendency and, therefore, gives a more balanced picture of the salary structure. The most acceptable method would be to report all three measures of central tendency and permit the audience to interpret the results. It is unfortunately true that few social researchers—let alone public relations practitioners or union representatives—report more than a single measure of central tendency. Even more unfortunate is the fact that some reports of research fail to specify exactly which measure of central tend-

TABLE 4.6
Measures of central tendency in a skewed distribution of annual salaries

Salary	
$100,000	
25,000	$\overline{X}$ = $18,000
10,000	
5,000	Mdn = $3,000
1,000	
1,000	Mo = $1,000
1,000	
1,000	

ency—the mode, median, or mean—was used to calculate the "average" amount or position in a group of scores. As shown by the foregoing illustration, a reasonable interpretation of findings may be impossible without such information.

It was noted earlier that some frequency distributions can be characterized as being bimodal, since they contain two points of maximum frequency. To properly describe bimodal distributions, it is generally useful to identify *both* modes; important aspects of such distributions can be obscured using either the median or the mean.

Consider the situation of a social researcher who conducted personal interviews with 26 lower-income respondents in order to determine their ideal conceptions of family size. Each respondent was asked: "Suppose you could decide exactly how large your family should be. Including all children and adults, how many people would you like to see in your ideal family?" As shown in Table 4.7, the results of this study indicated a wide range of family size preferences, from living alone (1) to living with many persons (10). Using either the mean or the median, we might conclude that the "average" respondent's ideal family contained about six persons ($\overline{X} = 5.58$; Mdn = 6). However, knowing that the distribution is bimodal, we see there were actually *two* ideal conceptions of family size represented in the group of respondents—one having a rather large number of persons (Mo = 8), and the other having only a few persons (Mo = 3).

The Research Objective

Until this point we have discussed the choice of a measure of central tendency in terms of the level of measurement and the shape of a distribution of scores. We now ask: what does the social researcher expect to do with his measure of central tendency? If he seeks a fast, simple, yet crude descriptive measure or if he is working with a bimodal distribution, he generally will employ the mode. In most situations encountered by the researcher, however, the mode is useful only as a preliminary indicator of central tendency, which can be quickly obtained from briefly scanning the data. If he seeks a pre-

TABLE 4.7
Ideal conceptions of family size among 26 lower-income respondents: a bimodal distribution

Ideal Family Size	f
10	1
9	2
8	6
7	3
6	2
5	1
4	2
3	6
2	2
1	1
	N = 26

cise measure of central tendency, the decision is usually between the median and the mean.

To describe a skewed distribution, the social researcher generally chooses the median, since (as was earlier noted) it tends to give a balanced picture of extreme scores. In addition, the median is sometimes used as a point in the distribution where the scores can be divided into two categories containing the same number of respondents. For instance, we might divide respondents at the median into two categories according to their family size preferences—those liking a small family versus those who like a large family.

For a precise measure of symmetrical distributions, the mean tends to be preferred over the median since the mean can be easily used in more advanced statistical analyses such as those introduced in subsequent chapters of the text. Moreover, the mean is more stable than the median, in that it varies less across different samples drawn from any given population. This advantage of the mean—although perhaps not yet understood or appreciated by the student—will become more apparent in subsequent discussions of the decision-making function of statistics (see Chapter 7).

OBTAINING THE MODE, MEDIAN, AND MEAN FROM A GROUPED FREQUENCY DISTRIBUTION

In a grouped frequency distribution, the mode is the midpoint of the class interval that has the greatest frequency. Following this definition, the mode for the distribution located in Table 4.8 is 72, since this is the midpoint of the interval that occurs most often (it occurs 17 times).

To locate the median of data grouped into a frequency distribution, we must (1) find the class interval containing the median and (2) interpolate.

Step 1—to locate the median interval, we first construct a cumulative frequency distribution as shown in the third column of Table 4.9. Starting with the interval that contains the lowest values (the youngest ages, 20–29), we add frequencies until arriving at that interval holding the case that divides the distribution into two

TABLE 4.8
Obtaining the mode from a grouped frequency distribution

Class Interval	Midpoint	f
95–99	97	3
90–94	92	2
85–89	87	4
80–84	82	7
75–79	77	12
70–74	72	17
65–69	67	12
60–64	62	5
55–59	57	5
50–54	52	4
		$N = \overline{71}$

TABLE 4.9
A grouped frequency distribution of age

Interval	f	cf
60–69	15	100
50–59	32	85
40–49	27	53
30–39	16	26
20–29	10	10
	$N = 100$	

equal parts, the middlemost score. In the present example, $N = 100$, and therefore we look for the fiftieth case ($N/2 = 100/2 = 50$). Moving up from the lowest interval, we see that 26 of the cases have ages of 39 or younger. We see also that the fiftieth case falls within the interval 40–49, since this is the class interval whose cumulative frequencies contain 53, or more than one-half of the cases. In other words, referring to the cumulative frequencies, the twenty-seventh through fifty-third cases are found within the interval 40–49. This is the median interval.

Step 2—to find the exact value of the median, we apply the formula

$$\text{Median} = \begin{array}{c}\text{lower limit}\\\text{of median}\\\text{interval}\end{array} + \left(\dfrac{\dfrac{N}{2} - \begin{array}{c}cf\ \text{below the}\\\text{lower limit of}\\\text{median interval}\end{array}}{f\ \text{in median interval}}\right) \begin{array}{c}\text{size of}\\\text{interval}\end{array}$$

For the data in Table 4.9, the median is determined as follows:

$$\text{Median} = 39.5 + \left(\frac{50 - 26}{27}\right) 10$$
$$= 39.5 + 8.89$$
$$= 48.39$$

To calculate the mean from a grouped frequency distribution, a modified version of the formula for a simple frequency distribution may be used (see Table 4.5). As shown below, the symbol X is no longer used to designate a score, but refers to the *midpoint of a class interval*. Therefore,

$$\overline{X} = \frac{\Sigma fX}{N}$$

where

$\overline{X}$ = the mean
X = the midpoint of a class interval
fX = a midpoint multiplied by the number of cases within its class interval
N = the total number of scores

We can illustrate the calculation of a mean from grouped data with reference to the following distribution:

Interval	f
17–19	1
14–16	2
11–13	3
8–10	5
5–7	4
2–4	2
	$N = 17$

STEP 1: Find the Midpoint of Each Class Interval

Interval	X = midpoint
17–19	18
14–16	15
11–13	12
8–10	9
5–7	6
2–4	3

STEP 2: Multiply Each Midpoint by the Number of Cases Within Its Interval and Obtain ΣfX

Interval	X = midpoint	f	fX
17–19	18	1	18
14–16	15	2	30
11–13	12	3	36
8–10	9	5	45
5–7	6	4	24
2–4	3	2	6
		$N = 17$	$\Sigma fX = 159$

STEP 3: "Plug the Result of Step 2 into the Formula for $\overline{X}$

$$\overline{X} = \frac{\Sigma fX}{N}$$

$$= \frac{159}{17}$$

$$= 9.35$$

SUMMARY

This chapter has introduced the three best known measures of central tendency, measures of what is "average" or "typical" of a set of data. The mode was defined as the category or score that occurs most often; the median was regarded as the middlemost point in a distribution; the mean was considered as the sum of a set of scores divided by the total number of scores in a set. These measures of central tendency were compared with regard to level of measurement, shape or form of their distribution, and the research objective. We can summarize those conditions for choosing among the three measures in the following way:

Mode:
1. level of measurement: nominal, ordinal, or interval
2. shape of distribution: most appropriate for bimodal
3. objective: fast, simple, but rough measure of central tendency

Median:
1. level of measurement: ordinal or interval
2. shape of distribution: most appropriate for highly skewed
3. objective: precise measure of central tendency, can sometimes be used for more advanced statistical operations or for splitting distributions into two categories (for example, high versus low)

Mean:
1. level of measurement: interval
2. shape of distribution: most appropriate for unimodal symmetrical
3. objective: precise measure of central tendency, can often be used for more advanced statistical operations including decision-making tests to be discussed in subsequent chapters of the text

PROBLEMS

1. The hourly wages of seven employees in a small company are $9, $8, $9, $4, $1, $6, and $3. Find (a) the modal hourly wage, (b) the median hourly wage, and (c) the mean hourly wage.
2. Suppose that the small company in Problem 1 above hired another employee at an hourly wage of $1, resulting in the following hourly wages: $9, $8, $9, $4, $1, $6, $3, and $1. Find (a) the modal hourly wage, (b) the median hourly wage, and (c) the mean hourly wage.
3. For the scores 205, 6, 5, 5, 5, 2, and 1, find (a) the mode, (b) the median, and (c) the mean. Which measure of central tendency would you *not* use to describe this set of scores? Why?
4. Six students in a sociology seminar were questioned by means of an interval-level measure regarding attitude toward vivisec-

tion. Their responses on the scale from 1 to 10 (higher scores indicate more favorable attitudes toward vivisection) were as follows: 5, 2, 6, 3, 1, and 1.

For the attitude scores above, find (a) the mode, (b) the median, and (c) the mean. On the whole, how favorable toward vivisection were these students?

5. For the scores 10, 12, 14, 8, 6, 7, 10, 10, find (a) the mode, (b) the median, and (c) the mean.

6. For the scores 3, 3, 4, 3, 1, 6, 5, 6, 6, 4, find (a) the mode, (b) the median, and (c) the mean.

7. For the scores 8, 8, 7, 9, 10, 5, 6, 8, 8, find (a) the mode, (b) the median, and (c) the mean.

8. For the scores 5, 4, 6, 6, 1, 3, find (a) the mode, (b) the median, and (c) the mean.

9. For the scores 8, 6, 10, 12, 1, 3, 4, 4, find (a) the mode, (b) the median, and (c) the mean.

10. For the scores 12, 12, 1, 12, 5, 6, 7, find (a) the mode, (b) the median, and (c) the mean.

11. What is the deviation of each of the following scores from a mean of 20.5? (a) $X = 20.5$; (b) $X = 33.0$; (c) $X = 15.0$; (d) $X = 21.0$.

12. What is the deviation of each of the following scores from a mean of 3.0? (a) $X = 4.0$; (b) $X = 2.5$; (c) $X = 6.3$; (d) $X = 3.0$.

13. What is the deviation of each of the following scores from a mean of 15? (a) $X = 22.5$; (b) $X = 3$; (c) $X = 15$; (d) $X = 10.5$.

14. The scores of attitudes toward Puerto Ricans for 31 students were arranged in the following frequency distribution (higher scores indicate more favorable attitudes toward Puerto Ricans):

Attitude Score	f
7	3
6	4
5	6
4	7
3	5
2	4
1	2
	$N = 31$

Find (a) the mode, (b) the median, and (c) the mean.

15. The 31 children enrolled in an urban elementary school third grade class were asked to indicate the number of their brothers and/or sisters living at home. The resultant data were arranged in the form of a frequency distribution as follows:

Number of Siblings	f
5	6
4	7
3	9
2	5
1	4
	N = 31

For this class of 31 children, find (a) the modal number of siblings, (b) the median number of siblings, and (c) the mean number of siblings.

16. For the following frequency distribution, find (a) the mode, (b) the median, and (c) the mean:

Score Value	f
10	3
9	4
8	6
7	8
6	9
5	7
4	5
3	2
2	1
1	1
	N = 46

17. For the following grouped frequency distribution, find (a) the mode, (b) the median, and (c) the mean:

Class Interval	f
20–24	2
15–19	4
10–14	8
5–9	5
	N = 19

18. For the following grouped frequency distribution, find (a) the mode, (b) the median, and (c) the mean:

Class Interval	f
90–99	16
80–89	17
70–79	15
60–69	3
50–59	2
40–49	3
	N = 56

19. For the following grouped frequency distribution, find (a) the mode, (b) the median, and (c) the mean:

Class Interval	f
17–19	2
14–16	3
11–13	6
8–10	5
5–7	1
	$N = 17$

TERMS TO REMEMBER

Central tendency
Mode
Median
Mean
Unimodal versus bimodal distributions
Deviation score

5 Measures of Variability

IN CHAPTER 4, we saw that the mode, median, and mean could be used to summarize in a single number what is "average" or "typical" of a distribution. When employed alone, however, any measure of central tendency yields only an incomplete picture of a set of data and, therefore, can mislead or distort as well as clarify.

In order to illustrate this possibility, consider that Honolulu, Hawaii, and Houston, Texas, have almost the same mean daily temperature of 75°. Can we, therefore, assume that the temperature is basically alike in both localities? Or is it not possible that one city is better suited for swimming and other out-of-door activities?

As shown in Figure 5.1, Honolulu's temperature varies only slightly throughout the year, usually ranging between 70° and 80°. By contrast, the temperature in Houston can differ seasonally from a low of about 40° in January to a high of close to 100° in July and August. Needless to say, Houston's beaches are not overcrowded year around!

Take another example: suppose that burglars and high school teachers in a particular city were found to have the same mean annual income of $8000. Would this finding *necessarily* indicate that the two income distributions are alike? On the contrary, they might be found to differ markedly in another important respect—namely, that teachers' incomes cluster closely together around $8000, whereas burglars' incomes are more widely scattered, reflecting greater opportunity for prison, unemployment, and poverty as well as for unusually great wealth.

It can be seen that we need, in addition to a measure of central tendency, an index of how the scores are scattered around the center of the distribution. In a word, we need a measure of what is commonly referred to as *variability* (also known as *spread, width,* or *dispersion*). Returning to an earlier example, we might say that the distribution of temperature in Houston, Texas has *greater variability* than the distribution of temperature in Honolulu, Hawaii. In the same way, we can say that the distribution of income among teachers has *less variability* than

FIGURE 5.1
*Differences in variability; the
distribution of temperature
in Honolulu and Houston
(approximate figures)*

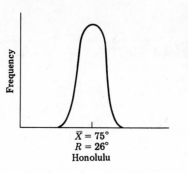
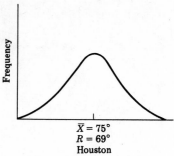

$\overline{X} = 75°$
$R = 26°$
Honolulu

$\overline{X} = 75°$
$R = 69°$
Houston

the distribution of income among burglars. This chapter discusses only the best known measures of variability: the *range,* the *mean deviation,* and the *standard deviation.*

THE RANGE

To get a quick but rough measure of variability, we might find what is known as the range R, the difference between the highest and lowest scores in a distribution. For instance, if Honolulu's hottest annual temperature was 88° and its coldest annual temperature was 62°, then the range of annual temperature in Honolulu would be 26° (88 − 62 = 26). If Houston's hottest day was 102° and its coldest day 33°, the range of temperature in Houston would be 69° (102 − 33 = 69).

The advantage of the range—its quick and easy computation—is also its most important disadvantage. That is, the range is totally dependent on only two score values, the largest and smallest cases in a given set of data. As a result, the range usually gives merely a crude index of the variability of a distribution. For instance, $R = 98$ in the data 2, 6, 7, 7, 10, 12, 13, 100 ($R = 100 − 2 = 98$), whereas $R = 12$ in the data 2, 6, 7, 7, 10, 12, 13, 14 ($R = 14 − 2 = 12$). Therefore, by changing only a *single* score (from 100 to 14), we caused the range to fluctuate sharply from 98 to 12. Any measure that is so much affected by the score of only one respondent cannot possibly give a precise idea as to variability and, at best, must be considered to be only a preliminary or very rough index.

**THE MEAN
DEVIATION**

In the previous chapter, the concept of deviation was defined as the distance of any given raw score from its mean. To find deviation, we were told to subtract the mean from any raw score ($x = X − \overline{X}$). If now we wish to obtain a measure of variability which takes into account every score in a distribution (rather than only two score values), we might take the absolute deviation (or distance) of each score from the mean of the distribution ($|x|$), add these deviations, and then divide this sum by the number of scores. The result would be the mean deviation. By formula,

$$MD = \frac{\Sigma|x|}{N}$$

where

MD = the mean deviation
$\Sigma|x|$ = the sum of the absolute deviations (disregarding plus and minus signs)
N = total number of scores

An important note: to get $\Sigma|x|$, we *must* ignore plus and minus signs and add absolute values. This is true because the sum of actual deviations (Σx)—deviations using signs to show direction whether above or below the mean—is always equal to zero. Plus and minus deviations cancel themselves out and therefore cannot be used to describe or compare the variability of distributions. By contrast, the sum of absolute deviations tends to become larger as the variability of a distribution increases.

We can now illustrate the step-by-step procedure for computing the mean deviation by considering the set of scores 9, 8, 6, 4, 2, and 1.

STEP 1: Find the Mean for the Distribution

X	
9	
8	$\bar{X} = \frac{\Sigma X}{N}$
6	
4	$= \frac{30}{6}$
2	
1	$= 5$
$\Sigma X = 30$	

STEP 2: Subtract the Mean from Each Raw Score and Add These Deviations (Ignoring Their Signs)

X	x
9	+4
8	+3
6	+1
4	−1
2	−3
1	−4
$\Sigma X = 30$	$\Sigma|x| = 16$

STEP 3: Divide $\Sigma|x|$ by N in Order to Control for the Number of Cases Involved

$$MD = \frac{\Sigma|x|}{N}$$

$$= \frac{16}{6}$$

$$= 2.67$$

Following the foregoing procedure, we see that the mean deviation is 2.67 for the set of scores 9, 8, 6, 4, 2, and 1. This indicates that, on the average, scores in this distribution deviate from the mean by 2.67 units.

To better understand the utility of mean deviation, let us turn to daily income distributions (a), (b), and (c) as located in Table 5.1. Notice first that the mean of each distribution is $20. Note also that there seem to be important differences in variability among the distributions—differences that can be detected with the help of the range and the mean deviation.

Let us first examine income distribution (a), in which all incomes are exactly alike. Since all scores in this distribution take on identical numerical values ($20), we can say that distribution (a) has no variability at all. Everyone earned the same amount of money on that day. As a result, the range is 0, and there is absolutely no deviation from the mean (MD = 0). Distributions (b) and (c) do contain variability. More specifically, distribution (b) has a range of 6 and a mean deviation of 1.71; distribution (c) has a range of 30 and a mean deviation of 8.57. We can therefore say that distribution (b) contains less variability than distribution (c)—the incomes in distribution (b) are more alike than are the incomes in distribution (c).

TABLE 5.1
Variability in distributions of daily income having the same mean ($20)

Distribution (a)		Distribution (b)		Distribution (c)	
X	$|x|$	X	$|x|$	X	$|x|$
$20	0	$23	+3	$35	+15
20	0	22	+2	30	+10
20	0	21	+1	25	+5
20	0	20	0	20	0
20	0	19	−1	15	−5
20	0	18	−2	10	−10
20	0	17	−3	5	−15
$\Sigma|x| = \overline{0}$		$\Sigma|x| = \overline{12}$		$\Sigma|x| = \overline{60}$	
$\overline{X}$ = $20		$\overline{X}$ = $20		$\overline{X}$ = $20	
R = $ 0		R = $ 6		R = $30	
MD = $ 0		MD = $ 1.71		MD = $ 8.75	
No variability		Some variability		Most variability	

THE STANDARD DEVIATION

For reasons that will soon become apparent, the mean deviation is no longer widely used by social researchers; it has been largely abandoned as a measure of variability in favor of its more effective "first cousin," the *standard deviation*. As we shall see, however, the mean deviation cannot be regarded as a waste of time, for, if nothing else, it provides us with a sound basis for understanding the nature of the standard deviation.

In a previous discussion, we learned that the mean deviation avoids the problem of negative numbers, which cancel out positive numbers by ignoring minus and plus signs and summing the absolute deviations from the mean. This procedure for creating a measure of variability has the distinct disadvantage that such absolute values are not always useful in more advanced statistical analyses (since they cannot easily be manipulated algebraically).

To overcome this problem and obtain a measure of variability which is more amenable to advanced statistical procedures, we might *square the actual deviations from the mean and add them together* (Σx^2). As illustrated in Table 5.2, this procedure would get rid of minus signs, since squared numbers are always positive.

Having added the squared deviations from the mean, we might *divide this sum by N in order to control for the number of scores involved and obtain what is known as the mean of these squared deviations.* (*Note:* You may recall that a similar procedure was followed to get the mean deviation when we divided $\Sigma|x|$ by N.) Continuing with the illustration in Table 5.2, we see that

$$\frac{\Sigma x^2}{N} = \frac{52}{6} = 8.67$$

One further problem arises. As the direct result of having squared deviations from the mean, the unit of measurement is changed, making our result 8.67 rather difficult to interpret. We have 8.67, but 8.67 units of what? In order to return to our original unit of measurement, then, we *take the square root of the mean of the squared deviations:*

$$\sqrt{\frac{\Sigma x^2}{N}} = \sqrt{8.67} = 2.95$$

TABLE 5.2
Squaring deviation scores to eliminate negative numbers: an illustration using data from Table 5.1

X	x	x^2
9	+4	16
8	+3	9
6	+1	1
4	−1	1
2	−3	9
1	−4	16
	$\Sigma x = 0$	$\Sigma x^2 = 52$

We now define the standard deviation as the result of the foregoing series of operations, that is, as *the square root of the mean of the squared deviations from the mean of a distribution.* Symbolized by SD or the lowercase Greek letter sigma σ,

$$\sigma = \sqrt{\frac{\Sigma x^2}{N}}$$

where

σ = the standard deviation
Σx^2 = the sum of the squared deviations from the mean
N = the total number of scores

To summarize, the procedure for computing the standard deviation does not differ much from the method we learned earlier to obtain the mean deviation. With reference to the present example, the following steps are carried out.

STEP 1: Find the Mean for the Distribution

X	
9	$\overline{X} = \dfrac{\Sigma X}{N}$
8	
6	$= \dfrac{30}{6}$
4	
2	$= 5$
1	
$\Sigma X = \overline{30}$	

STEP 2: Subtract the Mean from Each Raw Score to Get Deviation

X	x
9	+4
8	+3
6	+1
4	−1
2	−3
1	−4

STEP 3: Square Each Deviation Before Adding the Squared Deviations Together

X	x	x^2
9	+4	16
8	+3	9
6	+1	1
4	−1	1
2	−3	9
1	−4	16
		$\Sigma x^2 = \overline{52}$

STEP 4:   Divide by N and Get the Square Root of the Result

$$\sigma = \sqrt{\frac{\Sigma x^2}{N}}$$

$$= \sqrt{\frac{52}{6}}$$

$$= \sqrt{8.67}$$

$$= 2.95$$

We can now say that the standard deviation is 2.95 for the set of scores 9, 8, 6, 4, 2, and 1.

The Raw-Score Formula for SD

Until now, the formula $\sqrt{\Sigma x^2/N}$ has been used to compute the standard deviation. Especially if a calculating machine is available, there is an easier method for obtaining SD—a method which does not require getting deviations, but which works directly with the raw scores.

The raw-score formula is

$$\sigma = \sqrt{\frac{\Sigma X^2}{N} - \overline{X}^2}$$

where

σ = the standard deviation
ΣX^2 = the sum of the squared raw scores (important: each raw score is *first* squared and then these squared raw scores are added)
N = total number of scores
$\overline{X}^2$ = the mean squared

The step-by-step procedure for computing SD by the raw-score method can be illustrated by returning to the data in Table 5.2.

STEP 1:   Square Each Raw Score Before Adding the Squared Raw Scores Together

X	X^2
9	81
8	64
6	36
4	16
2	4
1	1
	$\Sigma X^2 = \overline{202}$

STEP 2:   Obtain the Mean and Square It

X
9
8
6
4
2
1

$$\bar{X} = \frac{\Sigma X}{N} = \frac{30}{6} = 5$$

$$\overline{X}^2 = 25$$

$\Sigma X = 30$

STEP 3: "Plug" the Results from Steps 1 and 2 into the Formula

$$\sigma = \sqrt{\frac{\Sigma X^2}{N} - \overline{X}^2}$$

$$= \sqrt{\frac{202}{6} - 25}$$

$$= \sqrt{33.67 - 25.00}$$

$$= \sqrt{8.67}$$

$$= 2.95$$

As shown above, applying the raw-score formula to the data in Table 5.2 yields exactly the same result as the original method.

Obtaining SD from a Simple Frequency Distribution

To obtain the standard deviation from data arranged in the form of a simple frequency distribution, we apply the formula

$$\sigma = \sqrt{\frac{\Sigma f X^2}{N} - \overline{X}^2}$$

To provide a step-by-step illustration, let us find the standard deviation for the following distribution:

Score Value	f
7	1
6	2
5	3
4	5
3	2
2	2
1	1
	$N = 16$

STEP 1: Multiply Each Score Value (X) by Its f to Obtain fX

X	f	fX
7	1	7
6	2	12
5	3	15
4	5	20
3	2	6
2	2	4
1	1	1

STEP 2: Multiply Each fX by X to Obtain fX^2 (Before Adding to Get ΣfX^2)

X	fX	fX^2
7	7	49
6	12	72
5	15	75
4	20	80
3	6	18
2	4	8
1	1	1
		$\Sigma fX^2 = 303$

STEP 3: Obtain the Mean and Square It

fX
7
12
15
20
6
4
1
$\Sigma fX = 65$

$$\overline{X} = \frac{\Sigma fX}{N}$$

$$= \frac{65}{16}$$

$$= 4.06$$

$$\overline{X}^2 = 16.48$$

STEP 4: "Plug" the Results from Steps 1, 2, and 3 into the Formula

$$\sigma = \sqrt{\frac{\Sigma fX^2}{N} - \overline{X}^2}$$

$$= \sqrt{\frac{303}{16} - 16.48}$$

$$= \sqrt{18.94 - 16.48}$$

$$= \sqrt{2.46}$$

$$= 1.57$$

The Meaning of the Standard Deviation

The series of steps required to compute the standard deviation can leave the student with an uneasy feeling as to the meaning of his result. For example, suppose we learn that $\sigma = 4$ in a particular distribution of scores. What is indicated by this number? Exactly what can we say now about that distribution which we could not have said before?

The following chapter will seek to clarify the full meaning of the standard deviation. For now, we note briefly that the standard deviation (like the mean deviation before it) represents the "average variability" in a distribution, since it measures the average of deviations from the mean. The procedures of squaring and taking the square root also enter the picture, but chiefly so as to eliminate minus signs and return to the more convenient unit of measurement, the raw-score unit.

We note also that the greater the variability around the mean of a distribution, the larger the standard deviation. Thus, $\sigma = 4.5$ indicates greater variability than $\sigma = 2.5$. For instance, the distribution of daily temperature in Houston, Texas, has a larger standard deviation than does the distribution of temperature for the same period in Honolulu, Hawaii.

If we wish to discuss the distance of a table from a living room wall, we might think in terms of yards or feet as a unit of measurement (for example, "The table in the living room is located 5 ft from this wall."). But how do we measure the width of a base line of a frequency polygon which contains the scores of a group of respondents arranged from low to high (in ascending order)? As a related matter, how do we come up with a method to find the distance between any raw score and its mean—a standardized method which permits comparisons between raw scores in the same distribution as well as between different distributions? If we were talking about tables, we might find that one table is 5 ft from the living room wall, while another table is 10 ft from the kitchen wall. In the concept of *feet,* we have a standard unit of measurement and, therefore, we can make such comparisons in a meaningful way. But how about comparing raw scores? For instance, can we always compare 85 on an English exam with 80 in German? Which grade is really higher? A little thought will show that it depends on how the other students in each class performed.

One method for giving a rough indicator of the width of a base line is the range, since it gives the distance between the highest and lowest scores along the base line. But the range cannot be effectively used to locate a score relative to its mean, since—aside from its other weaknesses—the range covers the entire width of the base line. By contrast, the size of the standard deviation is smaller than that of the range and usually covers far less than the entire width of the base line.

Just as we "lay off" a carpet in feet or yards, so we might "lay off" the base line in units of standard deviation (in sigma units). For instance, we might add the standard deviation to the value of the mean in order to find which raw score is located exactly one standard deviation (one sigma distance) from the mean. As shown in Figure 5.2, therefore, if $\overline{X} = 80$ and SD = 5, then the raw score 85 lies exactly one standard deviation *above* the mean $(80 + 5 = 85)$, a distance of $+1$ σ. This direction is "plus" because all deviations *above* the mean are positive; all deviations below the mean are "minus" or negative.

FIGURE 5.2
"Laying off" the base line in units of standard deviation when standard deviation (σ) is 5 and mean (X̄) is 80

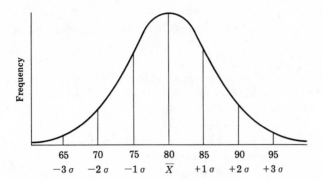

FIGURE 5.2
"Laying off" the base line in units of standard deviation when standard deviation (σ) is 5 and mean (X̄) is 80

We continue laying off the base line by adding the value of the standard deviation to the raw score 85. This procedure gives us the raw score 90, which lies exactly two standard deviations above the mean (85 + 5 = 90). Likewise, we add the standard deviation to the raw score 90 and obtain 95, which represents the raw score falling exactly three standard deviations above the mean. To continue the process below the mean, we subtract the standard deviation from the mean: we subtract 5 from 80, 5 from 75, and 5 from 70 to obtain $-1\,\sigma$, $-2\,\sigma$, and $-3\,\sigma$.

As illustrated in Figure 5.3, the process of laying off the base line in units of standard deviation is in many respects similar to measuring the distance between a table and the wall in units of feet. However, the analogy breaks down in at least one important respect: while feet and yards are of constant size (1 ft always equals 12 in.; 1 yd always equals 3 ft), the value of the standard deviation varies from distribution to distribution. Otherwise, we could not use the standard deviation as previously illustrated to compare distributions as to their variability (for example, SD = $5000 for the income distribution of high school teachers; SD = $15,000 for the income distribution of burglars). For this reason, we must calculate the size of the standard deviation for any distribution with which we happen to be working. Also as a result, it is usually more difficult to understand the standard deviation as opposed to feet or yards as a

FIGURE 5.3
Measuring the distance (a) between a table and a wall in units of feet and (b) between a raw score and a mean in units of standard deviation

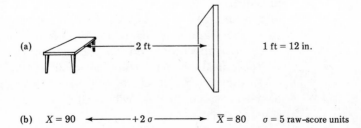

unit of measurement. We will return to this conception of the standard deviation in the following chapter.

COMPARING THE RANGE, MEAN DEVIATION, AND STANDARD DEVIATION

The range is regarded as merely a preliminary or rough index of the variability of a distribution. It is quick and simple to obtain, but not very reliable, and can be applied to interval or ordinal data.

The range does serve a useful purpose in connection with computations of the standard deviation. As illustrated in Figure 5.2, six standard deviations cover almost the entire distance from the highest to lowest score in a distribution ($-3\,\sigma$ to $+3\,\sigma$). This fact alone gives us a convenient method for estimating (but not computing) the standard deviation. Generally, the size of the standard deviation is approximately one-sixth of the size of the range. For instance, if the range is 36, then SD might be expected to fall close to 6; if the range is 6, SD will likely be close to 1.

This rule can take on considerable importance for the student who wishes to find out whether his result is anywhere in the vicinity of being correct. To take an extreme case, if $R = 10$ and our calculated SD $= 12$, we have made some sort of an error, since SD cannot be larger than the range. A note of caution: the one-sixth rule applies when we have a large number of scores. For a small number of cases, there will generally be a smaller number of standard deviations to cover the range of a distribution.

While the range is calculated from only two score values, both the standard deviation and the mean deviation take into account every score value in a distribution. Despite its relative stability, however, the mean deviation is no longer widely used in social research, since it cannot be employed in many advanced statistical analyses. By contrast, the standard deviation employs the mathematically acceptable procedure of clearing the signs (by squaring deviations) rather than ignoring them. As a result, the standard deviation has become the initial step for obtaining certain other statistical measures especially in the context of statistical decision making. We shall be exploring this characteristic of the standard deviation in detail in subsequent chapters, particularly in Chapters 6 and 7.

Despite its usefulness as a reliable measure of variability, the standard deviation also has its drawbacks. As compared with other measures of variability, the standard deviation tends to be difficult and time-consuming to calculate. However, this disadvantage is being more and more overcome by the increasing use of high-speed calculating machines and computers to perform statistical analyses. The standard deviation (like the mean deviation) also has the characteristic of being an interval-level measure and, therefore, cannot be used with nominal or ordinal data—data with which many social researchers often work.

OBTAINING THE RANGE, MEAN DEVIATION, AND STANDARD DEVIATION FROM GROUPED DATA

Whether working from grouped or ungrouped data, the range is always the difference between the highest and lowest scores. No special method or formula is necessary.

For purposes of illustrating the step-by-step procedure for obtaining the mean deviation for a grouped frequency distribution, consider the following grouped distribution:

Class Interval	f
17–19	1
14–16	2
11–13	3
8–10	5
5–7	4
2–4	2
	$N = 17$

STEP 1: Find the Midpoint of Each Class Interval

Interval	X = midpoint
17–19	18
14–16	15
11–13	12
8–10	9
5–7	6
2–4	3

STEP 2: Determine the Mean of the Distribution

X = midpoint	f	fX
18	1	18
15	2	30
12	3	36
9	5	45
6	4	24
3	2	6
		$\Sigma fX = 159$

$$\bar{X} = \frac{\Sigma fX}{N} = \frac{159}{17} = 9.35$$

STEP 3: Find the Deviation of Each Midpoint from the Mean

X = midpoint	$X - \bar{X} = |x|$
18	8.65
15	5.65
12	2.65
9	.35
6	3.35
3	6.35

Description

STEP 4: Multiply Each Deviation Score by the Frequency in the Respective Class Interval and Add These Products

Interval	f	$|x|$	$f|x|$
17–19	1	8.65	8.65
14–16	2	5.65	11.30
11–13	3	2.65	7.95
8–10	5	.35	1.75
5–7	4	3.35	13.40
2–4	2	6.35	12.70
	$N = 17$		$\Sigma f|x| = 55.75$

STEP 5: Divide by N

$$\text{MD} = \frac{\Sigma f|x|}{N}$$

$$= \frac{55.75}{17}$$

$$= 3.28$$

We arrive at a mean deviation of 3.28.

A raw-score formula can be used to calculate the standard deviation for a grouped frequency distribution. In formula terms,

$$\sigma = \sqrt{\frac{\Sigma f X^2}{N} - \overline{X}^2}$$

where

σ = the standard deviation
f = the frequency in a class interval
X = the midpoint of a class interval
N = the total number of scores
$\overline{X}^2$ = the mean squared

The step-by-step procedure for finding the standard deviation can be illustrated with reference to the grouped data:

Class Interval	f
17–19	1
14–16	2
11–13	3
8–10	5
5–7	4
2–4	2

STEP 1: Multiply Each Midpoint by the Frequency in the Class Interval and Add These Products

Class Interval	f	Midpoint (X)	fX
17–19	1	18	18
14–16	2	15	30
11–13	3	12	36
8–10	5	9	45
5–7	4	6	24
2–4	2	3	6
	$N = 17$		$\Sigma fX = 159$

STEP 2: Obtain the Mean and Square It

$$\overline{X} = \frac{\Sigma fX}{N}$$

$$= \frac{159}{17} \qquad \overline{X}^2 = 87.42$$

$$= 9.35$$

STEP 3: Multiply Each Midpoint by fX and Add These Products

Class Interval	f	Midpoint (X)	fX	fX²
17–19	1	18	18	324
14–16	2	15	30	450
11–13	3	12	36	432
8–10	5	9	45	405
5–7	4	6	24	144
2–4	2	3	6	18
				$\Sigma fX^2 = 1773$

STEP 4: "Plug" the Results of Steps 2 and 3 into the Formula

$$\sigma = \sqrt{\frac{\Sigma fX^2}{N} - \overline{X}^2}$$

$$= \sqrt{\frac{1773}{17} - 87.42}$$

$$= \sqrt{104.29 - 87.42}$$

$$= \sqrt{16.87}$$

$$= 4.11$$

The standard deviation turns out to be 4.11.

SUMMARY

In the present chapter, we have been introduced to the range, the mean deviation, and the standard deviation—three measures of variability or how the scores are scattered around the center of a distribution. The range was regarded as a quick but very rough indicator of variability, which can be found easily by taking the difference between the highest and lowest scores in a distribution. The mean deviation—the sum of the absolute deviations divided by

N—was treated as a mathematically inadequate measure of variability, but a sound basis for understanding the standard deviation, the square root of the mean of the squared deviations from the mean. In the standard deviation, we have a reliable, interval-level measure of variability, which can be utilized for more advanced descriptive and decision-making statistical operations. The full meaning of the standard deviation will be explored in subsequent discussions of the normal curve and generalizations from samples to populations.

PROBLEMS

1. The examination scores obtained by a group of 5 students are 7, 5, 3, 2, and 1 on a 10-point scale. For this set of scores, find (a) the range, (b) the mean deviation, and (c) the standard deviation.

2. On a scale designed to measure attitude toward racial segregation, two college classes scored as follows:

Class A	Class B
4	3
6	3
2	2
1	1
1	4
1	2

Compare the variability of attitudes toward racial segregation among the members of classes A and B by finding (a) the range of scores for each class, (b) the mean deviation of scores for each class, and (c) the standard deviation of scores for each class. Which class has greater variability of attitude scores?

3. For the set of scores 3, 5, 5, 4, 1, find (a) the range, (b) the mean deviation, and (c) the standard deviation.

4. For the set of scores 1, 6, 6, 3, 7, 4, 10, calculate the standard deviation.

5. For the set of scores 12, 12, 10, 9, 8, calculate the standard deviation.

6. Find the standard deviation for the following frequency distribution of scores:

X	f
5	3
4	5
3	6
2	2
1	2
	$N = 18$

7. Find the standard deviation for the following frequency distribution of scores:

X	f
7	2
6	3
5	5
4	7
3	4
2	3
1	1
	$N = 25$

8. Find the standard deviation for the following frequency distribution of scores:

X	f
10	2
9	5
8	8
7	7
6	4
5	3
	$N = 29$

9. For the following grouped frequency distribution of scores, find (a) the range, (b) the mean deviation, and (c) the standard deviation.

Class Interval	f
90–99	6
80–89	8
70–79	4
60–69	3
50–59	2
	$N = 23$

10. For the following grouped frequency distribution of scores, find (a) the range, (b) the mean deviation, and (c) the standard deviation:

Class Interval	f
17–19	2
14–16	3
11–13	6
8–10	5
5–7	1

11. For the following grouped frequency distribution of scores, find (a) the range, (b) the mean deviation, and (c) the standard deviation:

Class Interval	f
20–24	2
15–19	4
10–14	8
5–9	5
	$N = \overline{19}$

TERMS TO REMEMBER

Variability
Range
Mean deviation
Standard deviation

PART II

FROM DESCRIPTION
TO DECISION MAKING

6 The Normal Curve

IN PREVIOUS CHAPTERS, we saw that frequency distributions can take a variety of shapes or forms. Some are perfectly symmetrical or free of skewness; others are skewed either negatively or positively; still others have more than one "hump," and so on. Within this great diversity, there is one frequency distribution with which many of us are already familiar, if only from being graded by instructors on the "curve." This distribution, commonly known as the *normal curve,* is a theoretical or ideal model which was obtained from a mathematical equation, rather than from actually conducting research and gathering data.[1] However, the usefulness of the normal curve for the social researcher can be seen in its applications to actual research situations.

As we shall see in the present chapter, for example, the normal curve can be used for describing distributions of scores, interpreting the standard deviation, and making statements of probability. In subsequent chapters, we shall see that the normal curve is an essential ingredient of statistical decision making, whereby the social researcher generalizes his results from samples to populations. Before proceeding to a discussion of techniques of decision making, it is first necessary to gain an understanding of the properties of the normal curve.

---

[1] The normal curve can be constructed from the formula,

$$Y = \frac{N}{\sigma\sqrt{2\pi}} e^{-(X-\bar{X})^2/2\sigma^2}$$

where

    Y = the ordinate for a given value of X (its frequency of occurrence)
    π = 3.1416
    e = 2.7183

CHARACTERISTICS OF THE NORMAL CURVE

How can the normal curve be characterized? What are the properties that distinguish it from other distributions? As indicated in Figure 6.1, the normal curve is a type of smooth, symmetrical curve whose shape reminds many individuals of a bell and is thus widely known as the "bell-shaped curve." Perhaps the most outstanding feature of the normal curve is its *symmetry:* if we were to fold the curve at its highest point at the center, we would create two equal halves, each the mirror image of the other.

In addition, the normal curve is *unimodal,* having only one peak or point of maximum frequency—that point in the middle of the curve at which the mean, median, and mode coincide (the student may recall that the mean, median, and mode occur at different points in a skewed distribution—see Chapter 3). From the rounded central peak of the normal distribution, the curve falls off gradually at both tails, extending indefinitely in either direction and getting closer and closer to the base line without actually reaching it.

NORMAL CURVES: THE MODEL AND THE REAL WORLD

We might ask: to what extent do distributions of actual data (that is, the data collected by social researchers in the course of doing research) closely resemble or approximate the form of the normal curve? For illustrative purposes, let us imagine that all social, psychological, and physical phenomena were normally distributed. What would this hypothetical world be like?

So far as physical human characteristics are concerned, most adults would fall within the 5 to 6 ft range of height, with far fewer being either very short (less that 5 ft) or very tall (more than 6 ft). As shown in Figure 6.2, IQ would be equally predictable—the greatest proportion of IQ scores would fall between 90 and 110; we would see a gradual falling off of scores at either end with few "geniuses" who score higher than 140 and equally few who score lower than 60. Likewise, relatively few individuals would be regarded as political extremists, either of the right or of the left, while most would be considered politically moderate or "middle of the roaders." Finally, even the pattern of wear resulting from the flow of traffic in

FIGURE 6.1
The shape of the normal curve

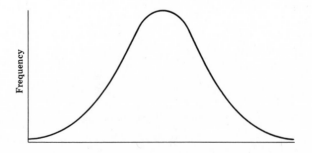

FIGURE 6.2
*Hypothetical distribution of
IQ scores*

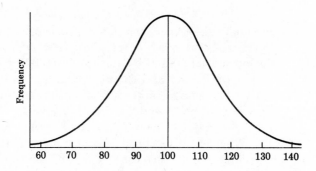

doorways would resemble the normal distribution—most wear would take place in the center of the doorway, whereas gradually decreasing amounts of wear would occur at either side.

    Some readers have, by this time, noticed that the hypothetical world of the normal curve does not differ radically from the "real world" in which we presently live. Phenomena such as height, IQ, political orientation, and wear in doorways do, in fact, seem to approximate the theoretical normal distribution. Because so many phenomena have this characteristic—because it occurs so frequently in nature (and for other reasons that will soon become apparent)—researchers in many fields have made extensive use of the normal curve by applying it to the data which they collect and analyze.

    But it should also be noted that some phenomena in social science, as elsewhere, simply do not conform to the theoretical notion of the normal distribution. Many distributions are skewed; others have more than one peak; some are symmetrical but not bell shaped. As a concrete example, let us consider the distribution of wealth throughout the world. It is well known that the "have-nots" greatly outnumber the "haves." Thus, as shown in Figure 6.3, the distribu-

FIGURE 6.3
The distribution of per capita income among the nations of the world (in U.S. dollars)

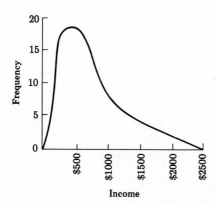

tion of wealth (as indicated by per capita income) is apparently extremely skewed, so that a small proportion of the world's population receives a large proportion of the world's income. Likewise, population specialists tell us that the United States has recently been a land of the young and the old. From an economic standpoint, this distribution of age represents a burden for a relatively small labor force made up of "middleaged" citizens providing for a disproportionately large number of retired as well as school-age dependents.

Where we have good reason to expect radical departures from normality—as in the case of age and income—the normal curve cannot be used as a model of the data we have obtained. Thus, it cannot be applied at will to all of the distributions encountered by the researcher, but must be used with a good deal of discretion. Fortunately, statisticians know that many phenomena of interest to the social researcher take the form of the normal curve.

THE AREA UNDER THE NORMAL CURVE

In order to employ the normal distribution in solving problems, we must acquaint ourselves with the area under the normal curve: *that area which lies between the curve and the base line containing 100 percent or all of the cases in any given normal distribution.* Figure 6.4 illustrates this characteristic.

We could enclose a portion of this total area by drawing lines from any two points on the base line up to the curve. For instance, using the mean as a point of departure, we could draw one line at $\overline{X}$ and another line at the point which is 1 SD (1 sigma distance) above $\overline{X}$. As illustrated in Figure 6.5, this shaded portion of the normal curve includes 34.13 percent of the total frequency.

In the same way, we can say that 47.72 percent of the cases under the normal curve lie between $\overline{X}$ and 2 SDs above $\overline{X}$, and that 49.87 percent lie between $\overline{X}$ and 3 SDs above $\overline{X}$ (see Figure 6.6).

As we shall see, *a constant proportion of the total area under the normal curve will lie between the mean and any given distance from $\overline{X}$ as measured in SD units.* This is true regardless of the mean and SD of the particular distribution and applies universally to all data which are normally distributed. Thus, the area under the normal curve between $\overline{X}$ and the point 1 SD above $\overline{X}$ *always* turns out to in-

FIGURE 6.4
The area under the normal curve

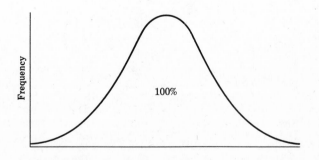

FIGURE 6.5
The percent of total area under the normal curve between X̄ and the point one standard deviation above X̄

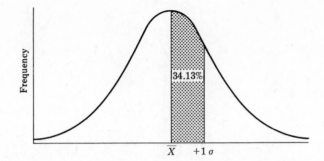

clude 34.13 percent of the total cases, whether we are discussing the distribution of height, intelligence, political orientation, or the pattern of wear in a doorway. The basic requirement, in each case, is only that we are working with a *normal* distribution of scores.

The symmetrical nature of the normal curve leads us to make another important point—namely, that *any given sigma distance above the mean contains the identical proportion of cases as the same sigma distance below the mean.* Thus, if 34.13 percent of the total area lie between the mean and 1 SD *above* X̄, then 34.13 percent of the total area also lie between the mean and 1 SD *below* X̄; if 47.72 percent lie between the mean and 2 SDs *above* X̄, then 47.72 percent lie between the mean and 2 SDs *below* X̄; if 49.87 percent lie between the mean and 3 SDs *above* X̄, then 49.87 percent also lie between the mean and 3 SDs *below* X̄. In other words, as illustrated in Figure 6.7, 68.26 percent of the total area of the normal curve (34.13 percent + 34.13 percent) fall between $-1\ \sigma$ and $+1\ \sigma$ from the mean; 95.44 percent of the area (47.72 percent + 47.72 percent) fall between $-2\ \sigma$ and $+2\ \sigma$ from the mean; 99.74 percent, or almost all, of the cases (49.87 percent + 49.87 percent) fall between $-3\ \sigma$ and $+3\ \sigma$ from the mean. It can be said, then, that 6 SDs include practically all of the cases (more than 99 percent) under any normal distribution.

FIGURE 6.6
The percent of total area under the normal curve between X̄ and the points two and three standard deviations from X̄

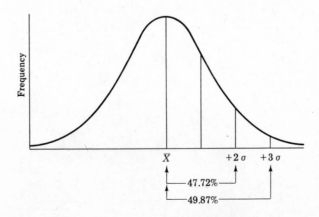

FIGURE 6.7

The percent of total area under the normal curve between −1σ and +1σ, −2σ and +2σ, and −3σ and +3σ

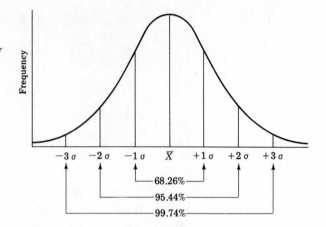

CLARIFYING THE STANDARD DEVIATION: AN ILLUSTRATION

An important function performed by the normal curve is that of interpreting and clarifying the meaning of the standard deviation. To understand how this function is carried out, let us examine what anthropologists tell us about sex differences in IQ. Despite the claims of male chauvinists, there is evidence that both males and females have mean IQ scores of approximately 100. Let us also say these IQ scores differ markedly in terms of variability around the mean. In particular, let us suppose that male IQs have greater *heterogeneity* than female IQs; that is, the distribution of male IQs contains a much larger percent of extreme scores representing very bright as well as very dull individuals, whereas the distribution of female IQs has a larger percent of scores located closer to the average, the point of maximum frequency at the center.

Because the standard deviation is a measure of variation, these sex differences in variability should be reflected in the value of the SD of each distribution of IQ scores. Thus, we might find that the SD is 10 for male IQs, but the SD is only 5 for female IQs.

Knowing the standard deviation of each set of IQ scores and assuming that each set is normally distributed, we could then esti-

FIGURE 6.8

A distribution of male IQ scores

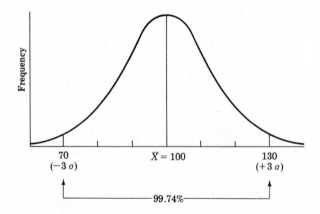

FIGURE 6.9
A distribution of female IQ scores

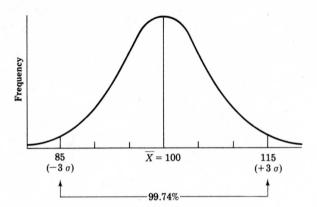

mate and compare the percent of males and females having any given range of IQ scores.

For instance, measuring the base line of the distribution of male IQ in SD units, we would know that 68.26 percent of male IQ scores fall between $-1\,\sigma$ and $+1\,\sigma$ from the mean. Since the standard deviation is always given in raw-score units and $\sigma = 10$, we would also know that these are points on the distribution at which IQ scores of 110 and 90 are located ($\overline{X} - \sigma = X$: $100 - 10 = 90$ and $100 + 10 = 110$). Thus, 68.25 percent of the males would have IQ scores that fall between 90 and 110.

Moving away from $\overline{X}$ and farther out from these points, we would find, as illustrated in Figure 6.8, that 99.74 percent, or practically all, of the males have IQ scores between 70 and 130 (between $-3\,\sigma$ and $+3\,\sigma$).

In the same manner, looking next at the distribution of female IQ scores as depicted in Figure 6.9, we see that 99.74 percent of these cases would fall between scores of 85 and 115 (between $-3\,\sigma$ and $+3\,\sigma$). In contrast to males, then, the distribution of female IQ scores could be regarded as being relatively *homogeneous,* having a smaller proportion of extreme scores in either direction. This difference is reflected in the comparative size of each SD and in the IQ scores falling between $-3\,\sigma$ and $+3\,\sigma$ from the mean.

USING TABLE B

In discussing the normal distribution, we have so far treated only those distances from the mean that are exact multiples of the standard deviation. That is, they were precisely 1, 2, or 3 SDs either above or below the mean. The question now arises: what must we do to determine the percent of cases for distances lying between any two ordinates? For instance, suppose we wish to determine the percent of total frequency that falls between the mean and, say, a raw score located 1.40 SDs above the mean. As illustrated in Figure 6.10, a raw score 1.40 SDs above the mean is obviously greater than 1 SD but less than 2 SDs from the mean. Thus, we know this dis-

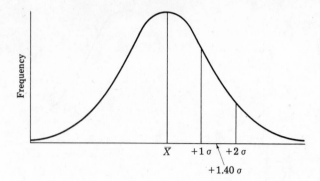

tance from the mean would include more than 34.13 percent but less that 47.72 percent of the total area under the normal curve.

To determine the *exact* percentage within this interval, we must employ Table B at the back of the text, which gives the percent under the normal curve between the mean and various sigma distances from the mean. These sigma distances (from 0.0 to 5.0) are shown in the left-hand column of Table B and have been given to one decimal place. The second decimal place has been given in the first or top row of the Table.

Notice that the symmetry of the normal curve makes it possible to give percentages for only one side of the mean, that is, only one-half of the curve (50 percent). Values in Table B represent either side. A portion of Table B has been reproduced below.

z	.00	.01	.02	.03	.04	.05	.06	.07	.08	.09
0.0	00.00	00.40	00.80	01.20	01.60	01.99	02.39	02.79	03.19	03.59
0.1	03.98	04.38	04.78	05.17	05.57	05.96	06.36	06.75	07.14	07.53
0.2	07.93	08.32	08.71	09.10	09.48	09.87	10.26	10.64	11.03	11.41
0.3	11.79	12.17	12.55	12.93	13.31	13.68	14.06	14.43	14.80	15.17
0.4	15.54	15.91	16.28	16.64	17.00	17.36	17.72	18.08	18.44	18.79

When learning to use and understand Table B, we might first attempt to locate the percent of cases between a sigma distance of 1.0 and the mean (the reason being that we already know that 34.13 percent of the total area fall between these points on the base line). Looking at Table B, we see it indeed indicates that exactly 34.13 percent of the total frequency fall between the mean and a sigma distance of 1.00. Likewise, we see that the sigma distance 2.00 includes exactly 47.72 percent of the total area under the curve, while the sigma distance 2.01 contains 47.78 percent of this total area.

STANDARD SCORES AND THE NORMAL CURVE

We are now prepared to find the percent of the total area under the normal curve associated with any given sigma distance from the mean. However, at least one more important question remains to be answered: how do we determine the sigma distance of any given raw score? That is, how do we go about translating our raw score—that score which we originally collected from our respondents—into units of standard deviation? If we wished to translate feet into yards, we would simply divide the number of feet by 3, since there are 3 ft in a yard. Likewise, if we were translating minutes into hours, we would divide the number of minutes by 60, since there are 60 min in every hour. In precisely the same manner, we can translate any given raw score into SD units by dividing the distance of the raw score from the mean by the SD. To illustrate, let us imagine a raw score of 6 from a distribution in which the mean is 3 and the SD is 2. Taking the difference between the raw score and the mean and obtaining a deviation score (6 − 3), we see that a raw score of 6 is 3 raw-score units above the mean. Dividing this raw-score distance by the SD of 2, we see that this raw score is 1.5 (one and one-half) standard deviations above the mean. In other words, the sigma distance of a raw score of 6 *in this particular distribution* is 1.5. We should note that regardless of the measurement situation, there are always 3 ft in a yard and 60 min in an hour. The constancy that marks these other standard measures is not shared by the standard deviation. It changes from one distribution to another. For this reason, we must know the standard deviation of a distribution either by calculating it, estimating it, or being given it by someone else, before we are able to translate any particular raw score into units of standard deviation.

The process that we have just illustrated—that of finding sigma distance from $\overline{X}$—yields a value called a *z score* or *standard score*, which indicates the *direction and degree that any given raw score deviates from the mean of a distribution on a scale of SD units* (notice that the left-hand column of Table B in the back of the book is labeled "*z*"). Thus, a *z* score of +1.4 indicates that the raw score lies 1.4 SDs (or almost $1\frac{1}{2}$ SDs) *above* the mean, while a *z* score of −2.1 means that the raw score falls slightly more than 2 SDs *below* the mean (see Figure 6.11).

We obtain a *z* score by finding the deviation score ($x = X - \overline{X}$) (which gives the distance of the raw score from the mean), and then dividing this raw-score deviation by σ.

FIGURE 6.11
The position of z = −2.1 and z = +1.4 in a normal distribution

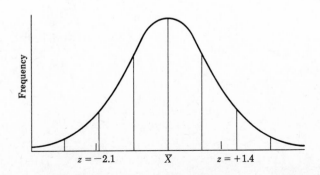

Computed by formula,

$$z = \frac{X - \overline{X}}{\sigma} \quad \text{or} \quad \frac{x}{\sigma}$$

where

x = a deviation score
σ = the standard deviation of a distribution
z = a standard score

Example 1

We are studying the distribution of annual income in a town in which the mean annual income is $5000 and the standard deviation is $1500. Assuming the distribution of annual income is normally distributed, we can translate the raw score from this distribution, $7000, into a standard score in the following manner:

$$z = \frac{7000 - 5000}{1500} = +1.33$$

Thus, an annual income of $7000 is 1.33 standard deviations above the mean annual income of $5000 (see Figure 6.12).

Example 2

We are working with a normal distribution of scores representing satisfaction with public housing among a group of project tenants (higher scores indicate greater satisfaction with public housing). Let us say this distribution has a mean of 10 and a standard deviation of 2. To determine how many standard deviations a score of 3 lies from the mean of 10, we obtain the difference between this score and the mean, that is,

$$\begin{aligned} x &= X - \overline{X} \\ &= 3 - 10 \\ &= -7 \end{aligned}$$

We then divide by the standard deviation:

$$\begin{aligned} z &= \frac{x}{\sigma} \\ &= -\tfrac{7}{2} \\ &= -3.5 \end{aligned}$$

Thus, as shown in Figure 6.13, a raw score 3 in this distribution of scores falls 3.5 standard deviations below the mean.

Note: If we know a z score and seek to obtain its raw-score equivalent, we use the formula

$$X = z\sigma + \overline{X}$$

For the present example,

$$\begin{aligned} X &= (-3.5)(2) + 10 \\ &= -7 + 10 \\ &= 3 \end{aligned}$$

FIGURE 6.12
*The position of z = 1.33 for
the raw score $7000*

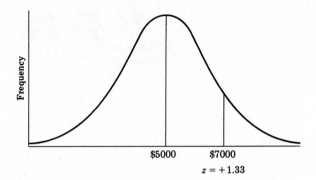

**PROBABILITY AND
THE NORMAL
CURVE**

As we shall now see, the normal curve can be used in conjunction with z scores and Table B to determine the probability of obtaining any raw score in a distribution. In the present context, the term *probability* refers to the relative frequency of occurrence of any given outcome or event; that is, *the probability associated with an event is the number of times that event can occur relative to the total number of times any event can occur.*

$$\text{The probability of an outcome or event} = \frac{\text{number of times the outcome or event can occur}}{\text{total number of times any outcome or event can occur}}$$

For example, if a room contains three men and seven women, the probability that the next person coming out of the room is a man would be 3 in 10:

$$\text{The probability of a man coming out next} = \frac{\text{number of men in the room}}{\text{total number of men and women in the room}} = \frac{3}{10}$$

In the same way, the probability of drawing a single card (let us say, the ace of spades) from a shuffled pack of 52 cards is 1 in 52, since the outcome "ace of spades" can occur only once out of the total number of times any outcome can occur, 52 cards. The probability of

FIGURE 6.13
*The position of z = −3.5 for
the raw score 3*

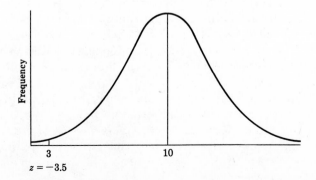

getting "heads" from an unbiased or perfectly balanced coin, which is flipped just once, is 1 in 2, since "heads" occurs once in the total number of times any outcome can occur, which is 2.

In the present context, the normal curve is a distribution in which it is possible to determine probabilities associated with various points along its base line. As noted earlier, the normal curve is a *frequency distribution* in which the total frequency under the curve equals 100 percent—it contains a central area surrounding the mean, where scores occur most frequently, and smaller areas toward either end, where there is a gradual flattening out and thus a smaller proportion of extremely high and low scores. In probability terms, then, we can say that probability decreases as we travel along the base line away from the mean in either direction. Thus, to say that 68.26 percent of the total frequency under the normal curve fall between $-1\,\sigma$ and $+1\,\sigma$ from the mean is to say that the probability is approximately 68 in 100 that any given raw score will fall within this interval. Similarly, to say that 95.44 percent of the total frequency under the normal curve fall between $-2\,\sigma$ and $+2\,\sigma$ from the mean is also to say that the probability is approximately 95 in 100 that any raw score will fall within this interval, and so on.

This is precisely the same concept of probability or *relative frequency* that we saw in operation when drawing a single card from an entire pack of cards or flipping a coin. Note, however, that the probabilities associated with areas under the normal curve are always given relative to 100 percent which is the entire area under the curve (for example, 68 in 100, 95 in 100, 99 in 100, and so on). For this reason, and to provide a standard way of looking at probability throughout this book, we shall be treating *probability as the number of times any given event could occur out of 100*. Thus, the probability of drawing the ace of spades from a shuffled deck is 1.92 in 100 ($\frac{1}{52}$) and of getting "heads" in one flip is 50 in 100 ($\frac{1}{2}$). Note, moreover, that probability is usually expressed in decimals as a proportion (P). For instance, we can say that $P = .50$ ($\frac{50}{100}$) of getting "heads" in one flip. Likewise, we can say that $P = .68$ ($\frac{68}{100}$) that any given raw score will fall between $-1\,\sigma$ and $+1\,\sigma$ under the normal curve.

Expressed as a proportion, *probability always ranges between 0 and 1*. The probability of an outcome is 0 when we are absolutely sure that it will not occur; the probability of an outcome is 1 when we are absolutely sure that it will occur. Social researchers are less than absolutely certain about anything! As a result, we might expect frequently to find a probability equal to .60, .25, or .05; but we would hardly ever expect to reduce probability to 0 or to increase it to 1.

An important characteristic of probability is found in the *addition rule* which states that *the probability of obtaining any one of several different outcomes equals the sum of their separate probabilities*. Suppose, for example, that we wish to find the probability of drawing *either* the ace of spades, the queen of diamonds, *or* the king

FIGURE 6.14
*The probability of obtaining
either the ace of spades, the
queen of diamonds, or the
king of hearts in a single
draw from a shuffled pack of
52 cards: an illustration of
the addition rule*

A♠	Probability of getting the ace of spades	$\frac{1}{52}$
 Q♦	Probability of getting the queen of diamonds	$\frac{1}{52}$
 K♥	Probability of getting the king of hearts +	$\frac{1}{52}$
	Probability of getting either the ace of spades, queen of diamonds, or king of hearts	$\frac{3}{52}$ $(P = .06)$

of hearts in a single draw from a well-shuffled pack of 52 cards. By adding their separate probabilities ($\frac{1}{52} + \frac{1}{52} + \frac{1}{52}$), we learn that the probability of obtaining any one of these cards in a single draw is equal to $\frac{3}{52}$ ($P = .06$). In other words, we have 6 chances in 100 to obtain either the ace of spaces, the queen of diamonds, or the king of hearts in a single draw. (See Figure 6.14.)

The addition rule always assumes that the outcomes being considered are *mutually exclusive*—that is, no two outcomes can occur simultaneously. For instance, no single card from a deck of 52 cards can be a spade, a diamond, *and* a heart at the same time. Similarly, a coin that is flipped just one time cannot possibly land both on its head *and* its tail.

Assuming mutually exclusive outcomes, we can say that the probability associated with all possible outcomes of an event always equals to 1. This indicates that some outcome must occur. If not heads, then tails; if not an ace, then a king, queen, jack, ten, and so on. In the flip of a coin, the probability of getting heads is equal to $\frac{1}{2}$ ($P = .50$). Of course, the probability of getting tails is also $\frac{1}{2}$ ($P = .50$). Adding together the probabilities for all possible outcomes, then, we learn that the probability of getting either heads or tails is equal to 1 ($\frac{1}{2} + \frac{1}{2} = 1$).

Another important property of probability occurs in the *multiplication rule* which focuses on the problem of obtaining two or more outcomes in sucessive order, one after another. The multiplication rule states that *the probability of obtaining a combination of mutually exclusive outcomes equals the product of their separate probabilities.* Rather than "either . . . or," the multiplication rule is addressed to "first, second, third."

For example, what is the probability of getting heads on two successive flips of a coin (heads on the first flip *and* heads on the second flip)? Since these outcomes are independent of one another, the outcome of the first flip does not influence the outcome of the second flip. On the first flip of the coin, the probability of obtaining heads equals $\frac{1}{2}$ ($P = .50$); on the second flip, the probability of obtaining heads still equals $\frac{1}{2}$ ($P = .50$). Therefore, the probability of getting heads on two successive flips of a coin is equal to $(\frac{1}{2})(\frac{1}{2}) = \frac{1}{4}$ (or $P = .25$). (See Figure 6.15.)

FIGURE 6.15
The probability of getting heads on two successive flips of a coin: an illustration of the multiplication rule

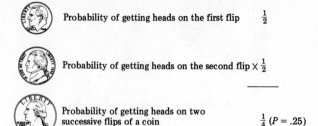

Probability of getting heads on the first flip $\frac{1}{2}$

Probability of getting heads on the second flip $\times \frac{1}{2}$

Probability of getting heads on two successive flips of a coin $\frac{1}{4}$ $(P = .25)$

To apply the foregoing conception of probability in relation to the normal distribution, let us return to an earlier example. We were then asked to translate into its *z*-score equivalent a raw score from a town's distribution of annual income, which we assumed approximated a normal curve. This distribution of income had a mean of $5000 with a SD of $1500.

By applying the *z*-score formula, we earlier learned that an annual income of $7000 was 1.33 SDs above the mean $5000; that is,

$$z = \frac{7000 - 5000}{1500} = +1.33$$

Let us now determine the probability of obtaining a score which lies between $5000, the mean, and $7000. In other words, what is the probability of randomly choosing, in just one attempt, a person from this town whose annual income falls between $5000 and $7000? The problem is graphically illustrated in Figure 6.16 (we are solving for the shaded area under the curve) and can be solved in two steps using the *z*-score formula and Table B at the back of the book.

STEP 1:   Translate the Raw Score ($7000) into a *z* Score

$$z = \frac{X - \overline{X}}{\sigma}$$

$$= \frac{7000 - 5000}{1500}$$

$$= +1.33$$

FIGURE 6.16
The portion of total area under the normal curve for which we seek to determine probability of occurrence

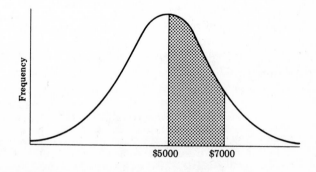

Thus, a raw score, $7000, is located 1.33 SDs above the mean.

STEP 2: Using Table B, Find the Percent of Total Frequency Under the Curve Falling Between the z Score ($z = +1.33$) and the Mean

In Table B, we find that 40.82 percent (41 percent) of the total population of this town fall (earn) between $5000 and $7000 (see Figure 6.17). Thus, by moving the decimal two places to the left, we see that the probability (rounded off) is 41 in 100: $P = .41$ that we would obtain an individual whose annual income lies between these figures.

In the foregoing example, we were asked to determine the probability associated with the distance between the mean and a particular sigma distance from it. Many times, however, we may wish to find the percent of area that lies at or *beyond* a particular raw score toward either tail of the distribution or to find the probability of obtaining these scores. For instance, in the present case, we might wish to learn the probability of obtaining an annual income of $7000 or *greater*.

This problem can be illustrated graphically as shown in Figure 6.18 (we are solving for the shaded area under the curve). In this case, we would follow Steps 1 and 2 above, thus obtaining the z score and finding the percent under the normal curve between $5000 and a $z = 1.33$ (from Table B). In the present case, however, we must go a step beyond and *subtract* the percentage obtained in Table B from 50 percent—that percent of the total area lying on either side of $\overline{X}$. This is true because *Table B always refers to the percent of area between a z score and the mean, never to the percent of area at or beyond a z score.*

Therefore, subtracting 40.82 percent from 50 percent, we learn that slightly more than 9 percent (9.18 percent) fall *at* or *beyond* $7000. In probability terms, we can say (by moving the decimal two places to the left) there are only slightly more than 9 chances in 100

FIGURE 6.17
The percent of total area under the normal curve between $\overline{X}$ = $5000 and z = +1.33

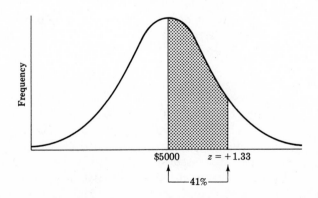

$5000   z = +1.33

└————41%————┘

FIGURE 6.18
*The portion of total area
under the normal curve for
which we seek to determine
probability of occurrence*

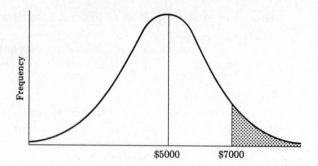

$(P = .09)$ that we would find an individual in this town whose income was $7000 or greater.

It was noted earlier that any given sigma distance above the mean contains the identical proportion of cases as the same sigma distance below the mean. For this reason, our procedure in finding probabilities associated with points below $\overline{X}$ is identical to that followed in the examples above.

For instance, the percent of total frequency between the z score -1.33 ($3000) and the mean is identical to the percent between the z score $+1.33$ ($7000) and the mean. Therefore, we know that $P = .41$ of obtaining an individual whose income falls between $3000 and $5000. Likewise, the percent of total frequency at or beyond -1.33 ($3000 or less) equals that at or beyond $+1.33$ ($7000 or more). Thus, we know $P = .09$ that we shall obtain someone from the town with an annual income of $3000 or less.

We can use the addition rule to find the probability of obtaining more than a single portion of the area under the normal curve. For instance, we have already determined that $P = .09$ for incomes of $3000 or less and for incomes of $7000 or more. To find the probability of obtaining *either* $3000 or less *or* $7000 or more, we simply add their separate probabilities as follows:

$P = .09 + .09$
$\quad = .18$

In a similar way, we can find the probability of obtaining someone whose income falls between $3000 and $7000 by adding the probabilities associated with z scores of 1.33 on either side of the mean. Therefore,

$P = .41 + .41$
$\quad = .82$

Notice that .82 + .18 equals 1, representing all possible outcomes under the normal curve.

The application of the multiplication rule to the normal curve can be illustrated by finding the probability of obtaining four individuals whose incomes are $7000 or greater. We already know that $P = .09$ associated with finding a single individual whose income is at least $7000. Therefore,

$$P = (.09)(.09)(.09)(.09)$$
$$= (.09)^4$$
$$= .00007$$

Applying the multiplication rule, we see that the probability of obtaining four individuals with incomes of $7000 or more is 7 chances in 100,000.

SUMMARY

This chapter sought to relate properties of the theoretical normal distribution to the "real-world" problems of social research. Thus, it was shown that the area under the normal curve can be used to interpret the standard deviation and make statements of probability. The importance of the normal distribution will become more apparent in subsequent chapters of the text.

PROBLEMS

1. In any normal distribution of scores, what percent of the total area falls (a) between -1 SD and $+1$ SD; (b) between -2 SD and $+2$ SD; (c) between -3 SD and $+3$ SD?
2. Given a normal distribution of raw scores in which $\overline{X} = 7.5$ and SD $= 1.3$, express each of the following raw scores as a z score: (a) $X = 8.0$; (b) $X = 6.0$; (c) $X = 5.3$; (d) $X = 10.2$; (e) $X = 7.5$; (f) $X = 8.5$; (g) $X = 11$.
3. Given a normal distribution of daily income in which $\overline{X} = \$10.50$ and SD $= \$2.80$, express each of the following incomes as a z score: (a) $X = \$8.40$; (b) $X = \$11.00$; (c) $X = \$13.20$; (d) $X = \$5.00$; (e) $X = \$15.00$; (f) $X = \$11.50$; (g) $X = \$9.00$.
4. For the income distribution in Problem 3 above, determine (a) the percent of respondents who earn a daily income of $15 or more; (b) the probability of locating a respondent whose daily income is $15 or more; (c) the percent of respondents who earn between $10 and $10.50; (d) the probability of locating a respondent whose income falls between $10 and $10.50; (e) the probability of locating a respondent whose income is $10 or less; (f) the probability of locating a respondent whose income is *either* $10 or less *or* $11 or more; (g) the probability of locating two respondents whose income is $10 or less.
5. Given a normal distribution of raw scores in which $\overline{X} = 80$ and SD $= 7.5$. determine (a) the percent of respondents who scored 60 or less; (b) the probability of locating a respondent who scored 60 or less; (c) the percent of respondents who scored between 80 and 90; (d) the probability of locating a respondent who scored between 80 and 90; (e) the percent of respondents who scored 85 or more; (f) the probability of locating a respondent who scored 85 or more; (g) the probability of locating a respondent who scored *either* 70 or less *or* 90 or more; (h) the probability of obtaining three respondents who scored 90 or more.

TERMS TO
REMEMBER

Normal curve
Area under the normal curve
z score (standard score)
Probability
Addition rule
Multiplication rule

7 Samples and Populations

THE SOCIAL RESEARCHER generally seeks to draw conclusions about large numbers of individuals. For instance, he or she might study the 220,000,000 citizens of the United States, the 1000 members of a particular labor union, the 10,000 black Americans who are living in a southern town, or the 45,000 students currently enrolled in an eastern university.

Until this point, we have been pretending that the social researcher investigates the entire group that he or she tries to understand. Known as a *population* or *universe,* this group consists of a set of individuals who share at least one characteristic, whether common citizenship, membership in a voluntary association, ethnicity, college enrollment, or the like. Thus, we might speak about the population of the United States, the population of labor union members, the population of black Americans residing in a southern town, or the population of university students.

Since social researchers operate with limited time, energy, and economic resources, they rarely study each and every member of the population in which they are interested. Instead, researchers study only a *sample*—a smaller number of individuals taken from some population. Through the sampling process, social researchers seek to generalize from a sample (a small group) to the entire population from which it was taken (a larger group).

The process of sampling is an integral part of everyday life. How else would we gain much information about other people than by sampling those around us? For example, we might casually discuss political issues with other students in order to find out where students generally stand with respect to their political opinions; we might attempt to determine how our classmates are studying for a particular examination by contacting only a few members of the class beforehand; we might even invest in the stock market after finding that a small sample of our associates have made money in a similar fashion.

SAMPLING METHODS

The social researcher's methods of sampling are usually more thoughtful and systematic than those of everyday life. He or she is centrally concerned with whether sample members are representative enough of the entire population to permit making accurate generalizations about that population. In order to make such inferences, the researcher selects an appropriate sampling method according to whether or not each and every member of the population has an equal chance of being drawn into the sample. If every population member is given an equal chance of sample selection, a *random* sampling method is being used; otherwise, a nonrandom type is employed.

Nonrandom Samples

The most popular nonrandom sampling method, *accidental* sampling, differs least from our everyday sampling procedures since it is based exclusively on what is convenient for the researcher. That is, the researcher simply includes the most convenient cases in his sample and excludes the inconvenient cases from his sample. Most students can recall at least a few instances when an instructor, who is doing research, has asked all of the students in his class to take part in an experiment or to fill out a questionnaire. The popularity of this form of accidental sampling in psychology has provoked some observers to view psychology as "the science of the college sophomore," since so many college students are the subjects for research.

Another nonrandom type is *quota* sampling. In this sampling procedure, diverse characteristics of a population, such as age, sex, social class, or ethnicity, are sampled in the proportions which they occupy in the population. Suppose, for instance, that we were asked to draw a quota sample from the students attending a university, where 42 percent were females and 58 percent were males. Using this method, interviewers are given a quota of students to locate, so that only 42 percent of the sample consists of females and 58 percent of the sample consists of males. The same percentages are included in the sample as are represented in the larger population. If the total sample size is 200, then 84 female students and 116 male students are selected.

A third variety of nonrandom sample is known as *judgment* or *purposive* sampling. The basic idea involved in this type is that logic, common sense, or sound judgment can be used to select a sample that is representative of a larger population. For instance, to draw a judgment sample of magazines that reflect middle-class values, we might, on an intuitive level, select *Reader's Digest, People,* or *Parade,* since articles from these titles *seem* to depict what most middle-class Americans desire (for example, the fulfillment of the American dream, economic success, and the like). In a similar way, state districts, which have traditionally voted for the winning candidates for state office, might be polled in an effort to predict the outcome of a current state election.

Random Samples

As previously noted, random sampling gives each and every member of the population an equal chance of being selected for the sample. This characteristic of random sampling indicates that every member of the population must be identified before the random sample is drawn, a requirement usually fulfilled by obtaining a list which includes each and every population member. A little thought will suggest that getting such a complete list of the population members is not always an easy task, especially if one is studying a large and diverse population. To take a relatively easy example, where could we get a *complete* list of students currently enrolled in a large university? Those social researchers who have tried will attest to its difficulty. For a more laborious assignment, try finding a list of every resident in a large city. How can we be certain of identifying everyone, even those residents who do not wish to be identified?

The basic type of random sample, *simple random* sampling, can be obtained by a process not unlike that of the now familiar technique of putting everyone's name on separate slips of paper and, while blindfolded, drawing only a few names from a hat. This procedure ideally gives every population member an equal chance for sample selection since one, and only one, slip per person is included. For several reasons (including the fact that the researcher would need an extremely large hat), the social researcher attempting to take a random sample usually does not draw names from hats. Instead, he uses a *table of random numbers* such as Table H located at the back of the text. A portion of a table of random numbers has been reproduced below.

			Column Number																	
	1	2	3	4	5	6	7	8	9	10	11	12	13	14	15	16	17	18	19	20
1	2	3	1	5	7	5	4	8	5	9	0	1	8	3	7	2	5	9	9	3
2	6	2	4	9	7	0	8	8	6	9	5	2	3	0	3	6	7	4	4	0
3	0	4	5	5	5	0	4	3	1	0	5	3	7	4	3	5	0	8	9	0
4	1	1	8	3	7	4	4	1	0	9	6	2	2	1	3	4	3	1	4	8
5	1	6	0	3	5	0	3	2	4	0	4	3	6	2	2	2	3	5	0	0

(Row Number on left side)

A table of random numbers is constructed so as to generate a series of numbers having no particular pattern or order. As a result, the process of using a table of random numbers yields an unbiased sample similar to that produced by putting slips of paper in a hat and drawing names while blindfolded.

To draw a simple random sample by means of a table of random numbers, the social researcher first obtains his list of the population and assigns a unique identifying number to each and every member. For instance, if he is conducting research on the 500 students enrolled in "Introduction to Sociology," he might secure a list of students from the instructor and give each student a number from 001 to 500. Having prepared the list, he proceeds to draw the members of his sample from a table of random numbers. Let us say the researcher seeks to draw a sample of 50 students to represent the 500 members of a class population. He might enter the random numbers table at any number (with eyes closed, for example) and move in any direction taking appropriate numbers until he has selected the 50 sample members. Looking at a portion of the random numbers table above, we might arbitrarily start at the intersection of column 1 and row 3, moving from left to right to take every number that comes up between 001 and 500. The first numbers to appear at column 1 and row 3 are 0, 4, and 5. Therefore, student number 045 is the first population member to be chosen for the sample. Continuing from left to right, we see that 4, 3, and 1 come up next, so that student number 431 is selected. This process is continued until all 50 sample members have been taken. A note to the student: in using the table of random numbers, always disregard numbers that come up a second time or are higher than needed.

All random sample methods are actually variations of the simple random sampling procedure just illustrated. For instance, with *systematic* sampling, a table of random numbers is not required, since a list of population members is sampled by fixed intervals. Employing systematic sampling, then, every *n*th member of a population is included in a sample of that population. To illustrate, in drawing a sample from the population of 10,000 public housing tenants, we might arrange a list of tenants, take every tenth name on the list, and come up with a sample of 1000 tenants.

The advantage of systematic sampling is that a table of random numbers is not required. As a result, this method is less time-consuming than the simple random procedure, especially for sampling from large populations. On the negative side, taking a systematic sample assumes that position on a list of population members does not influence randomness. If this assumption is not taken seriously, the result may be to overselect certain population members, while underselecting others. This can happen, for instance, when houses are systematically sampled from a list in which corner houses (which are generally more expensive than other houses on the block) occupy a fixed position, or when the names in a telephone directory are sampled by fixed intervals so that names associated with certain ethnic ties are underselected.

Another variation on simple random sampling, the *stratified* sample, involves dividing the population into more homogeneous subgroups or *strata* from which simple random samples are then taken. Suppose, for instance, we wish to study the acceptance of various birth control devices among the population of a certain city. Since attitudes toward birth control vary by religion and socioeconomic status, we might stratify our population on these variables, thereby forming more homogeneous subgroups with respect to the acceptance of birth control. More specifically, say we could identify Catholics, Protestants, and Jews as well as upper-class, middle-class, and lower-class members of the population. Our stratification procedure might yield the following subgroups or strata:

> upper-class Protestants
> middle-class Protestants
> lower-class Protestants
> upper-class Catholics
> middle-class Catholics
> lower-class Catholics
> upper-class Jews
> middle-class Jews
> lower-class Jews

Having identified our strata, we proceed to take a simple random sample from each subgroup or stratum (for example, from lower-class Protestants, from middle-class Catholics, and so on), until we have sampled the entire population. That is, each stratum is treated for sampling purposes as a complete population, and simple random sampling is applied. Specifically, each member of a stratum is given an identifying number, listed, and sampled by means of a table of random numbers. As a final step in the procedure, the selected members of each subgroup or stratum are combined in order to produce a sample of the entire population.

Stratification is based on the idea that a homogeneous group requires a smaller sample than does a heterogeneous group. For instance, studying individuals who are walking on a downtown street corner probably requires a larger sample than studying individuals living in a middle-class suburb. One can usually find individuals walking downtown who have any combination of characteristics. By contrast, persons living in a middle-class suburb are generally more alike with respect to education, income, political orientation, family size, attitude toward work, to mention only a few characteristics.

On the surface, random stratified samples bear a striking resemblance to the nonrandom quota method as previously discussed, since both procedures usually require the inclusion of sample characteristics in the exact proportions which they contribute to the population. Therefore, if 32 percent of our sample is made up of middle-class Protestants, then exactly 32 percent of our sample must be drawn from middle-class Protestants; in the same way, if 11 percent of our population consists of lower-class Jews, then 11 percent of our sample must be similarly constituted, and so on. In the context of stratified sampling, an exception arises when a particular stratum is disproportionately well represented in the sample, making possible a more intensive subanalysis of that group.

Such an occasion might arise, for example, when black Americans who constitute a small proportion of a given population are "oversampled" in an effort to examine their characteristics more closely.

Despite their surface similarities, quota and stratified samples are essentially different. While members of quota samples are taken by whatever method is chosen by the investigator, members of stratified samples are always selected on a *random* basis, generally by means of a table of random numbers applied to a complete list of the population members.

Before leaving the topic of sampling methods, let us examine the nature of an especially popular form of random sampling known as the *cluster* method. Such samples are widely used to reduce the costs of large surveys, in which interviewers must be sent to scattered localities and much traveling is required. Employing the cluster method, at least two levels of sampling are carried out:

1. the *primary sampling unit* or cluster, which is some well-delineated area considered to include characteristics found in the entire population (for example, a state, census tract, city block, and so on), and
2. the sample members within each cluster.

For illustrative purposes, imagine we wanted to interview a representative sample of individuals living in a large area of the city. Drawing a simple random, systematic, or stratified sample of respondents scattered over a wide area would entail a good deal of traveling, not to mention time and money. By means of cluster sampling, however, we would limit our interviewing to those individuals who are located within relatively few clusters. We might begin, for example, by treating the city block as our primary sampling unit or cluster. We might proceed, then, by obtaining a list of all city blocks within the area, from which we take a simple random sample of blocks. Having drawn our sample of city blocks, we might select the individual respondents (or households) on each block by the same simple random method. More specifically, all individuals (or households) on each of the selected blocks are listed and, with the help of a table of random numbers, a sample of respondents from each block is chosen. Using the cluster method, any given interviewer locates one of the selected city blocks and contacts more than one respondent who lives there.

On a much wider scale, the same cluster procedure can be applied to nationwide surveys by treating counties, states, or parishes as the primary sampling units to be initially selected and by interviewing a simple random sample of each of the chosen counties, states, or parishes. In this way, interviewers need not cover each and every county, state, or parish, but only a much smaller number of such areas that have been randomly selected for inclusion.

SAMPLING ERROR

Throughout the remainder of the text, we shall be careful to distinguish between characteristics of the samples we actually study and populations to which we hope to generalize. In order to make this distinction in our statistical procedures, we can therefore no longer use the same symbols to signify the mean and the standard deviation of both sample and population. Instead, we must employ different symbols depending on whether we are referring to sample or population characteristics. With reference to the mean, we shall always symbolize the mean of a *sample* as $\overline{X}$ and the mean of a *population* as M. With reference to the standard deviation, we shall symbolize the standard deviation of a *sample* as s and the standard deviation of its *population* as σ.

The social researcher typically tries to obtain a sample which is representative of the larger population in which he has an interest. Since random samples give each and every population member the same chance for sample selection, they are in the long run more representative of population characteristics than their nonrandom counterparts. As discussed briefly in Chapter 1, however, by chance alone, we can *always* expect some difference between a sample, random or otherwise, and the population from which it is drawn. $\overline{X}$ will almost never be exactly the same as M; s will hardly ever be exactly the same as σ. Known as *sampling error,* this difference results regardless of how well the sampling plan has been designed and carried out, under the researcher's best intentions, where no cheating occurs and no mistakes have been made.

To illustrate the operation of sampling error, let us now turn to Table 7.1, which contains the population of 20 final examination grades and 3 samples A, B, and C drawn at random from this population (each taken with the aid of a table of random numbers). As expected, the population mean ($M = 71.55$) is not arithmetically identical with any of the three sample means; similarly, there are differences among the sample means themselves.

SAMPLING DISTRIBUTION OF MEANS

Given the presence of sampling error, the student may wonder how it is possible *ever* to generalize from a sample to a larger population. To come up with a reasonable answer, let us consider the work of a hypothetical social researcher studying radio listening among the one million residents of a mid-western city. To save time and money, he interviews only a sample taken at random from the entire population of residents. He draws 500 residents by means of a table of random numbers, asks each sample member, "How many minutes do you usually listen to the radio daily?" and finds the time spent listening ranges from 0 to 240 min. As shown in Figure 7.1, the mean time spent listening in a sample of 500 residents is 101.55 min.

It turns out that our hypothetical social researcher is mildly

TABLE 7.1
A population and three random samples of final examination grades

Population			Sample A	Sample B	Sample C
70	80	93	96	40	72
86	85	90	99	86	96
56	52	67	56	56	49
40	78	57	52	67	56
89	49	48	303	249	273
99	72	30			
96	94	1431	$\overline{X} = 75.75$	$\overline{X} = 62.25$	$\overline{X} = 68.25$
	$M = 71.55$				

eccentric, having a notable fondness for drawing samples from populations. So intense is his enthusiasm for sampling that he continues to draw many additional samples of 500 residents each and to calculate radio listening time for the members of each sample. This procedure continues until our eccentric researcher has drawn 98 *samples* containing 500 residents *each*. In the process of drawing 98 random samples, he actually studies 49,000 respondents (500 × 98 = 49,000).

Let us assume, as shown in Figure 7.2, that the entire population of our midwestern city has a mean radio listening time of 99.75 min. Also as illustrated in Figure 7.2, let us suppose that the samples taken by our eccentric social researcher yield means which range between 89 and 111 min. In line with our previous discussion, this could easily happen simply on the basis of sampling error.

Frequency distributions of *raw scores* can be obtained from both samples and populations. In a similar way, we can construct a *sampling distribution of means,* a frequency distribution of a large number of random sample *means* which have been drawn from the same population. Table 7.2 presents the 98 sample means collected by our eccentric social researcher in the form of a sampling distribution. As when working with a distribution of raw scores, the means in Table 7.2 have been arranged in consecutive order from high to low and their frequency of occurrence indicated in an adjacent column.

FIGURE 7.1
The mean listening time for a random sample taken from a hypothetical population

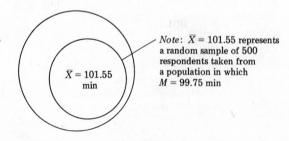

$\overline{X} = 101.55$ min

Note: $\overline{X} = 101.55$ represents a random sample of 500 respondents taken from a population in which $M = 99.75$ min

FIGURE 7.2
The mean listening time in 98 random samples taken from a hypothetical population in which M = 99.75 *min*

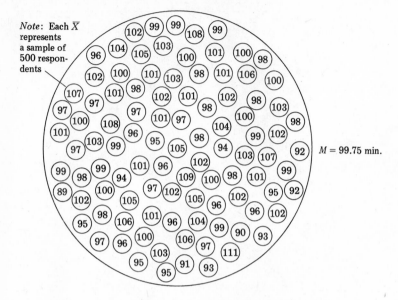

Note: Each $\overline{X}$ represents a sample of 500 respondents

M = 99.75 min.

Characteristics of a Sampling Distribution of Means

Until this point, we have not come directly to grips with the problem of generalizing from samples to populations. The theoretical model known as the sampling distribution of means (as illustrated by the 98 sample means obtained by our eccentric social researcher) has certain properties, which give to it an important role in the sampling process. Before moving on to the procedure for making generalizations from samples to populations, we must first examine the characteristics of a sampling distribution of means:

1. *The sampling distribution of means approximates a normal curve.* As graphically illustrated in Figure 7.3(a), by arranging the sample means from Table 7.2 in a frequency polygon, we end up with the form of the normal distribution. This is true of all sampling distributions of means regardless of the shape of the distribution of raw scores in the population from which the means are drawn.[1]

2. *The mean of a sampling distribution of means ("the mean of means") is equal to the true population mean.* If we take a large number of random sample means from the same population and find the mean of all sample means, we will have the value of the true population mean. Therefore, as shown in Figure 7.3, the mean of the sampling distribution of means (a) is the same as the mean of the population from which it was drawn (b). They can be regarded as interchangeable values.

3. *The standard deviation of a sampling distribution of means is smaller than the standard deviation of the population.*

[1] This assumes that we have drawn large random samples of equal size from a given population of raw scores.

TABLE 7.2
Sampling distribution of means (radio listening) for 98 random samples

Mean	f
111 min	1
110	1
109	1
108	2
107	2
106	3
105	4
104	5
103	6
102	8
101	9
100	9
99	9
98	8
97	7
96	6
95	5
94	4
93	3
92	2
91	1
90	1
89 min	1
	$N = \overline{98}$

As illustrated in Figure 7.3, the variability of the sampling distribution is always smaller than the variability of the entire population. This is true because we take mean scores (rather than the range of raw scores that make up those means), thereby eliminating extreme raw-score values. For instance, the mean alienation score 100 can be obtained from the raw scores 60, 90, 110, and 140. $(60 + 90 + 110 + 140 = \frac{400}{4} = 100.)$ Plotting raw scores, we include values ranging from 60 to 140. Plotting the mean score, however, we obviously reduce the occurrence of such extreme score values to a single value 100. As a result, we expect to obtain a smaller standard deviation when a number of mean scores are taken together and plotted.

FIGURE 7.3
Frequency polygons of (a) the sampling distribution of means from Table 7.2 and (b) the population from which these means were taken

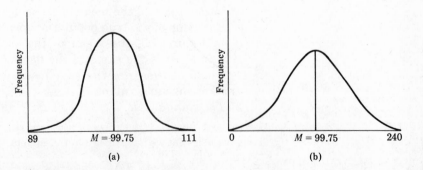

The Sampling Distribution of Means as a Normal Curve

As indicated in Chapter 6, if we define probability in terms of frequency of occurrence, then the normal curve can be regarded as a probability distribution (we can say that probability decreases as we travel along the base line away from the mean in either direction).

With this notion, we can find the probability of obtaining various raw scores in a distribution, given a certain mean and standard deviation. For instance, to find the probability associated with obtaining someone with an annual income between $5000 and $7000 in a population having a mean income of $5000 and a standard deviation of $1500, we translate the raw score $7000 into a z score (+1.33) and go to Table B in the back of the text to get the percent of total frequency falling between the z score 1.33 and the mean. This area contains 40.82 percent of the raw scores. Thus, $P = .41$ rounded off that we will find an individual whose annual income lies between $5000 and $7000. If we want the probability of finding someone whose income is $7000 or more, we must go a step beyond and subtract the percent obtained in Table B from 50 percent—that percentage of the area which lies on either side of the mean. Subtracting 40.82 percent from 50 percent, we learn that 9.18 percent falls at or beyond $7000. Therefore, moving the decimal two places to the left, we can say $P = .09$ (9 chances in 100) that we would find an individual whose income is $7000 or greater.

In the present context, we are no longer interested in obtaining probabilities associated with a distribution of *raw scores.* Instead, we find ourselves working with a distribution of *sample means,* which have been drawn from the total population of scores, and we wish to make probability statements about those sample means.

As illustrated in Figure 7.4, since the sampling distribution of means takes the form of the normal curve, we can say that probability decreases as we move farther away from the mean of means (true population mean). This makes sense because, as the student may re-

FIGURE 7.4
The sampling distribution of means as a probability distribution

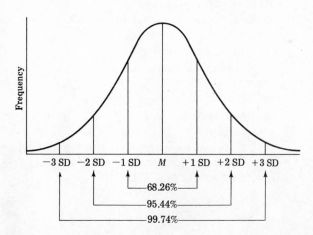

FIGURE 7.5
*The probability associated
with obtaining a sample
mean of $14,000 or less, if
the true population mean is
$20,000 and the standard
deviation is $2,600*

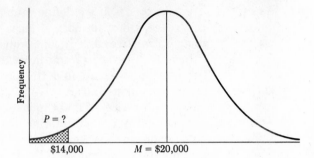

call, the sampling distribution is a product of chance differences
among sample means (sampling error). For this reason, we would
expect by chance and chance alone that most sample means will fall
close to the value of the true population mean, while relatively few
sample means will fall far from it.

Figure 7.4 indicates that about 68 percent of the sample means
in a sampling distribution fall between -1 SD and $+1$ SD from the
mean of means (true population mean). In probability terms, we can
say that $P = .68$ of any given sample mean falling within this interval. In the same say, we can say the probability is about .95 (95
chances out of 100) that any sample mean falls between -2 SD and
$+2$ SD from the mean of means, and so on.

Since the sampling distribution takes the form of the normal
curve, we are also able to use z scores and Table B to get the probability of obtaining any sample mean, not just those that are exact
multiples of the standard deviation. Given a mean of means and
standard deviation of the sampling distribution, the process is identical to that used in the previous chapter for a distribution of raw
scores. Only the names have been changed.

Imagine, for example, that a certain university claims its
alumni earn an average (M) annual income of $20,000. We have
reason to question the legitimacy of this claim and decide to test it
out on a random sample of 100 alumni. In the process, we get a
sample mean of only $14,000. We now ask: how probable is it that
we would get a mean of $14,000 or less if the true population mean
is actually $20,000? Has the university told the truth? Or is this only
an attempt to propagandize the public in order to increase enroll-
ments or endowments? Figure 7.5 illustrates the area for which we
seek a solution.

Suppose we know that the standard deviation of the sampling
distribution is $2600. Following the standard procedure, we trans-
late the sample mean $14,000 into a z score as follows:

$$Z = \frac{\overline{X} - M}{\sigma_{\overline{X}}} = \frac{14,000 - 20,000}{2600} = -2.31$$

where

$\overline{X}$ = a sample mean in the distribution

M = the mean of means (equal to the university's claim as to the true population mean)

$\sigma_{\overline{X}}$ = standard deviation of the sampling distribution of means

The result of the above procedure is to tell us that a sample mean of \$14,000 lies exactly 2.31 standard deviations below the claimed true population mean of \$20,000. Going to Table B in the back of the text, we see that 48.96 percent of the sample means fall between \$14,000 and \$20,000. Subtracting from 50 percent gives us the percent of the distribution that represents sample means of \$14,000 or less, if the true population mean is \$20,000. This figure is 1.04 percent (50 percent − 48.96 percent = 1.04 percent). Therefore, the probability is .01 rounded off (1 chance out of 100) of getting a sample mean of \$14,000 or less, when the true population mean is \$20,000. With such a small probability of being wrong, we can say with some confidence that the true population mean is *not* actually \$20,000. It is doubtful whether the university's report of alumni annual income represents more than bad propaganda.

STANDARD ERROR OF THE MEAN

Up until now, we have pretended that the social researcher actually has first-hand information about the sampling distribution of means. We have acted as though he, like the eccentric researcher, really has collected data on a large number of sample means, which were randomly drawn from some population. If so, it would be a simple enough task to make generalizations about the population, since the mean of means takes on a value that is equal to the true population mean.

In actual practice, the social researcher rarely collects data on more than one or two samples, from which he still expects to generalize to an entire population. Drawing a sampling distribution of means requires the same effort as it might take to study each and every population member. As a result, the social researcher does not have actual knowledge as to the mean of means or the standard deviation of the sampling distribution. However, he does have a good method for *estimating* the standard deviation of the sampling distribution of means on the basis of the data collected on a single sample. This estimate is known as the *standard error of the mean* and is symbolized by $\sigma_{\overline{X}}$.[2] By formula,

[2] In many texts, the standard error of the mean based on the population standard deviation and symbolized by $\sigma_{\overline{X}}$ is distinguished from the estimated standard error of the mean based on the sample standard deviation and symbolized by $s_{\overline{X}}$. Without measuring the entire population, however, the value of the population standard deviation is not known and must be estimated. For the sake of simplicity, we have therefore chosen to bypass the foregoing distinction and instead to introduce a single formula for the standard error of the mean, symbolized by $\sigma_{\overline{X}}$ and based on sample data.

$$\sigma_{\bar{X}} = \frac{s}{\sqrt{N-1}}$$

where

$\sigma_{\bar{X}}$ = standard error of the mean (an estimate of the standard deviation of a sampling distribution of means)

s = the standard deviation of a *sample*

N = the total number of scores in a *sample*

To illustrate, if the standard deviation of a sample of ten respondents is 2.5, then

$$\sigma_{\bar{X}} = \frac{2.5}{\sqrt{10-1}}$$

$$= \frac{2.5}{3.0}$$

$$= .83$$

As noted above, the social researcher who investigates only one or two samples cannot know the mean of means, the value of which equals the true population mean. He only has his obtained sample mean, which differs from the true population mean as the result of sampling error. But have we not come full circle to our original position? How is it possible to estimate the true population mean from a single sample mean, especially in light of such inevitable differences between samples and populations?

We have, in fact, traveled quite some distance from our original position. Having discussed the nature of the sampling distribution of means, we are now prepared to estimate the value of a population mean. With the aid of the standard error of the mean, we can find *the range of mean values within which our true population mean is likely to fall. We can also estimate the probability that our population mean actually falls within that range of mean values.* This is the concept of the *confidence interval.*

CONFIDENCE INTERVALS

In order to explore the procedure for finding a confidence interval, let us extend an earlier example. Suppose that a researcher's random sample of 100 alumni of a certain university yields a mean annual income of $14,000. Since his data come from only a random sample and not the entire population of alumni, we cannot be sure that the reported mean income is actually reflective of this population of university alumni. As we have already seen, sampling error is, after all, the inevitable product of sampling from populations.

We do know, however, that 68.26 percent of all random sample means in the sampling distribution of means will fall between -1 SD and $+1$ SD from the true population mean. By estimating the standard deviation of the sampling distribution ($\sigma_{\bar{X}} = \$2,000$) and

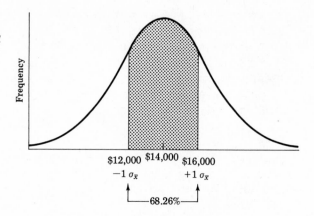

using our sample mean $14,000 as an estimate of the population mean, we can establish the range within which there are 68 chances out of 100 (rounded off) that the true population mean will fall. Known as the *68 percent confidence interval,* this range of mean incomes is graphically illustrated in Figure 7.6.

    The 68 percent confidence interval can be obtained in the following manner:

$$\text{the 68\% confidence interval} = \overline{X} \pm \sigma_{\bar{x}}$$

where

    $\overline{X}$ = a sample mean
    $\sigma_{\bar{x}}$ = the standard error of the mean

Applying the above formula to the problem at hand:

$$\begin{aligned}
\text{the 68\% confidence interval} &= \$14,000 \pm \$2,000 \\
&= \$12,000 \longleftrightarrow \$16,000
\end{aligned}$$

The social researcher, therefore, reports he is *68 percent confident* that the population mean income among these university alumni is $14,000, give or take $2,000. In other words, there are 68 chances in 100 ($P = .68$) that the true population mean actually falls within a range between $12,000 and $16,000 ($14,000 − $2,000 = $12,000; $14,000 + $2,000 = $16,000). This estimate is made despite sampling error, although within a range of error (plus or minus $2,000) and at a specified level of confidence (68 percent).

    Confidence intervals can be constructed for any level of probability. Most social researchers are not confident enough to estimate a population mean knowing there are only 68 chances out of 100 of being correct (68 out of every 100 sample means fall within the interval between $12,000 and $16,000). As a result, it has become a matter of convention to use a *wider,* less precise confidence interval having a *better probability* of making an accurate estimate of the population mean. Such a standard is found in the *95 percent confidence interval,* whereby the population mean is estimated, knowing there are 95 chances out of 100 of being right; there are 5 chances out of 100 of being wrong (95 out of every 100 sample means fall within the interval). Even using the 95 percent confidence interval, however, it must always be kept firmly in mind that the researcher's sample mean could be one of those five sample means which fall outside of the established interval. In statistical decision making, one never knows for certain.

How do we go about finding the 95 percent confidence interval? We already know that 95.44 percent of the sample means in a sampling distribution lie between -2 SD and $+2$ SD from the mean of means. Going to Table B, we can make the statement that 1.96 standard deviations in both directions cover exactly 95 percent of the sample means (47.50 percent on either side of the mean of means). In order to find the 95 percent confidence interval, we must first multiply the standard error of the mean by 1.96 (the interval is 1.96 units of $\sigma_{\bar{X}}$ in either direction from the mean). Therefore,

the 95% confidence interval $= \overline{X} \pm (1.96)\sigma_{\bar{X}}$

where

$\overline{X}$ = a sample mean
σ_X = the standard error of the mean

If we apply the 95 percent confidence interval to our estimate of the mean income among university alumni, we see

the 95% confidence interval $= \$14,000 \pm (1.96) \$2,000$
$= \$14,000 \pm \$3,920$
$= \$10,080 \longleftrightarrow \$17,920$

Conclusion: We are 95 percent confident that the true population mean falls between \$10,080 and \$17,920.

Let us summarize the step-by-step procedure for obtaining the 95 percent confidence interval in the following random sample of raw scores:

$\overline{X}$
1
5
2
3
4
1
2
2
4
3

STEP 1: Find the Mean of the Sample

X
1
5
2
3
4
1
2
2
4
3
$\Sigma X = 27$

$$\bar{X} = \frac{\Sigma X}{N}$$

$$= \frac{27}{10}$$

$$= 2.7$$

STEP 2: Obtain the Standard Deviation of the Sample

X	X^2
1	1
5	25
2	4
3	9
4	16
1	1
2	4
2	4
4	16
3	9
	$\Sigma X^2 = 89$

$$s = \sqrt{\frac{\Sigma X^2}{N} - \bar{X}^2}$$

$$= \sqrt{\tfrac{89}{10} - (2.7)^2}$$

$$= \sqrt{8.9 - 7.29}$$

$$= \sqrt{1.61}$$

$$= 1.27$$

STEP 3: Obtain the Standard Error of the Mean

$$\sigma_{\bar{X}} = \frac{s}{\sqrt{N-1}}$$

$$= \frac{1.27}{\sqrt{10-1}}$$

$$= \frac{1.27}{3}$$

$$= .42$$

STEP 4: Multiply the Standard Error of the Mean by 1.96

the 95% confidence interval $= \bar{X} \pm (1.96)\, \sigma_{\bar{X}}$
$$= 2.7 \pm (1.96)\,(.42)$$
$$= 2.7 \pm .82$$

STEP 5: Add and Subtract This Product from the Sample Mean in Order to Find the Range of Mean Scores Within Which the Population Mean Falls

$$\text{the 95\% confidence interval} = 2.7 \pm .82$$
$$= 1.88 \longleftrightarrow 3.52$$

We can be 95 percent confident that the true population mean lies between 1.88 and 3.52.[3]

An even more stringent confidence interval is the *99 percent confidence interval*. From Table B in the back of the text, we see that the *z* score 2.58 represents 49.50 percent of the area on either side of the curve. Doubling this amount yields 99 percent of the area under the curve; 99 percent of the sample means fall into that interval. In probability terms, 99 out of every 100 sample means fall between -2.58 SD and $+2.58$ SD from the mean. Conversely, only 1 out of every 100 means falls outside of the interval. By formula,

$$\text{the 99\% confidence interval} = \overline{X} \pm (2.58)\, \sigma_{\overline{X}}$$

where

$\overline{X}$ = a sample mean
$\sigma_{\overline{X}}$ = the standard error of the mean

With regard to our estimate of the mean income among university alumni:

$$\text{the 99\% confidence interval} = \$14{,}000 \pm (2.58)\, \$2{,}000$$
$$= \$14{,}000 \pm \$5{,}160$$
$$= \$8{,}840 \longleftrightarrow \$19{,}160$$

We have determined with 99 percent confidence that the true population mean falls somewhere between $8,840 and $19,160.

The student should note that the 99 percent confidence interval consists of a wider band ($8,840 to $19,160) than does the 95 percent confidence interval ($10,080 to $17,920). The 99 percent interval encompasses more of the total area under the normal curve and therefore a larger number of sample means. This wider band of mean scores gives us greater confidence that we have accurately estimated the true population mean. Only a single sample mean in every 100 lies outside of the interval. On the other hand, by increasing our confidence from 95 percent to 99 percent, we have also sacrificed a degree of precision in pinpointing the population mean. Holding sample size constant, the social researcher must choose between greater precision or greater confidence that he is correct.

To summarize the step-by-step procedure for finding the 99 percent confidence interval, let us reexamine the random sample of scores:

---

[3] For illustrative purposes, a small sample size was employed. In actual practice, however, a researcher who uses the foregoing procedure for finding a confidence interval must work with at least 30 cases in order to meet the assumption of normality in the sampling distribution of means (see the discussion of the *t* ratio in Chapter 8).

$\overline{X}$
1
5
2
3
4
1
2
2
4
3

STEP 1: Find the Mean of the Sample

1
5
2
3
4
1
2
2
4
3

$$\overline{X} = \frac{\Sigma X}{N}$$

$$= \frac{27}{10}$$

$$= 2.7$$

$\Sigma X = \overline{27}$

STEP 2: Obtain the Standard Deviation of the Sample

X	X^2
1	1
5	25
2	4
3	9
4	16
1	1
2	4
2	4
4	16
3	9
	$\Sigma X^2 = \overline{89}$

$$s = \sqrt{\frac{\Sigma X^2}{N} - \overline{X}^2}$$

$$= \sqrt{\frac{89}{10} - (2.7)^2}$$

$$= \sqrt{8.9 - 7.29}$$

$$= \sqrt{1.61}$$

$$= 1.27$$

STEP 3: Obtain the Standard Error of the Mean

$$\sigma_{\overline{X}} = \frac{s}{\sqrt{N-1}}$$

$$= \frac{1.27}{\sqrt{10-1}}$$

$$= \frac{1.27}{3}$$

$$= .42$$

STEP 4: Multiply the Standard Error of the Mean by 2.58

$$\text{the 99\% confidence interval} = \overline{X} \pm (2.58)\,\sigma_{\overline{X}}$$
$$= 2.7 \pm (2.58)\,(.42)$$
$$= 2.7 \pm 1.08$$

STEP 5: Add and Subtract This Product from the Sample Mean in Order to Find the Range of Mean Scores Within Which the Population Mean Falls

$$\text{the 99\% confidence interval} = 2.7 \pm 1.08$$
$$= 1.62 \longleftrightarrow 3.78$$

We are 99 percent confident that the true population mean falls between 1.62 and 3.78.

ESTIMATING PROPORTIONS

Thus far, we have focused on procedures for estimating population means. The social researcher often seeks to come up with an estimate of a population *proportion* strictly on the basis of a proportion he obtains in a random sample. A familiar circumstance is the pollster whose data suggest that a certain proportion of the vote will go to a particular political issue or candidate for office. When a pollster reports that 45 percent of the vote will be in favor of a certain candidate, he does so with the realization that he is less than 100 percent certain. In general, he is 95 percent or 99 percent confident that his estimated proportion falls within the range of proportions (for example, between 40 percent and 50 percent).

We estimate proportions by the procedure that we have just used to estimate means. All statistics—including means and proportions—have their sampling distributions. Just as we earlier found the standard error of the mean, we can now find the *standard error of the proportion*. By formula,

$$\sigma_P = \sqrt{\frac{P(1 - P)}{N}}$$

where

σ_P = standard error of the proportion (an estimate of the standard deviation of the sampling distribution of proportions)

P = a sample proportion

N = total number in the sample

For illustrative purposes, let us say 45 percent of a random sample of 100 college students report they are in favor of the legalization of all drugs. The standard error of the proportion would be

$$\sigma_P = \sqrt{\frac{.45(.55)}{100}}$$
$$= \sqrt{\frac{.2475}{100}}$$
$$= \sqrt{.0025}$$
$$= .05$$

To find the 95 percent confidence interval, we multiply the standard error of the proportion by 1.96 and add and subtract this product to and from the sample proportion:

the 95% confidence interval $= P \pm (1.96)\, \sigma_P$

where

P = a sample proportion
σ_P = standard error of the proportion

If we seek to estimate the proportion of college students in favor of the legalization of drugs,

$$\text{the 95\% confidence interval} = .45 \pm (1.96)\,.05$$
$$= .45 \pm .098$$
$$= .35 \longleftrightarrow .55$$

We are 95 percent confident that the true population proportion is neither smaller than .35 nor larger than .55. More specifically, somewhere between 35 percent and 55 percent of this population of college students are in favor of the legalization of all drugs. There is a 5 percent chance we are wrong; 5 times out of 100 such confidence intervals will not contain the true population proportion.

Let us summarize the procedure for estimating a proportion by means of the 95 percent confidence interval. Assume that the sample proportion from which our estimate is to be made turns out to be .40 (40 percent of the 100 cases fall into this category).

STEP 1: Obtain the Standard Error of the Proportion

$$\sigma_P = \sqrt{\frac{P(1 - P)}{N}}$$
$$= \sqrt{\frac{.40(.60)}{100}}$$
$$= \sqrt{\frac{.24}{100}}$$
$$= \sqrt{.0024}$$
$$= .049$$

STEP 2: Multiply the Standard Error of the Proportion by 1.96

$$\text{the 95\% confidence interval} = P \pm (1.96)\sigma_P$$
$$= .40 \pm (1.96)\,(.049)$$
$$= .40 \pm .096$$

STEP 3: Add and Subtract This Product from the Sample Proportion in Order to Find the Range of Proportions Within Which the Population Proportion Falls

$$\text{the 95\% confidence interval} = .40 \pm .096$$
$$= .30 \longleftrightarrow .50$$

We can say with 95 percent confidence that the true population proportion falls between .30 and .50.

SUMMARY

This chapter has explored the key concepts and procedures related to generalizing from samples to populations. Both random and nonrandom sampling methods were presented. It was pointed out that sampling error—the inevitable difference between samples and populations—occurs despite a well-designed and executed sampling plan. As a result of sampling error, we can discuss the characteristics of the sampling distribution of means, a distribution which forms a normal curve and whose standard deviation can be estimated with the aid of the standard error of the mean. Armed with such information, we can construct confidence intervals for means (or proportions) within which we have confidence (95 percent or 99 percent) that the true population mean (or proportion) actually falls. In this way, we are able to make generalizations from a sample to a population.

PROBLEMS

1. Find the standard error of the mean with the following sample of 30 scores:

3	5
3	3
2	3
1	2
5	2
4	3
5	2
1	4
6	6
3	1
2	1
1	3
1	4
2	3
3	4

2. With the sample mean in Problem 1 above, find (a) the 95 percent confidence interval, and (b) the 99 percent confidence interval.

3. Find the standard error of the mean with the following sample of 34 scores:

10	1
4	8
10	7
5	5
5	6
6	10
7	6
3	8
5	7
4	7
4	6
5	5
6	5
6	4
7	3
5	4
8	5

4. With the sample mean in Problem 3 above, find (a) the 95 percent confidence interval, and (b) the 99 percent confidence interval.

5. Find the standard error of the mean with the following sample of 32 scores:

4	4
2	3
5	6
6	6
1	7
1	1
7	5
8	7
7	8
8	8
8	4
2	5
6	3
5	2
6	6
4	5

6. With the sample mean in Problem 5 above, find (a) the 95 percent confidence interval, and (b) the 99 percent confidence interval.

7. In order to estimate the proportion of students on a particular campus who favor the abolition of fraternities, a social researcher interviewed a random sample of 50 students from the college population. He found that 57 percent of the sample favored getting

rid of fraternities (sample proportion = .57). With this information, (a) find the standard error of the proportion, and (b) construct a 95 percent confidence interval.

8. Given a sample size of 150 and a sample proportion of .32, (a) find the standard error of the proportion, and (b) construct a 95 percent confidence interval.

9. Given a sample size of 200 and a sample proportion of .25, (a) find the standard error of the proportion, and (b) construct a 95 percent confidence interval.

TERMS TO
REMEMBER

Population (universe)
Sample
Nonrandom sample
   Accidental
   Quota
   Judgment or purposive
Random sample
   Simple random sample
   Systematic sample
   Stratified sample
   Cluster sample
Table of random numbers
Primary sampling unit
Sampling error
Sampling distribution of means
Standard error of the mean
Confidence interval
95 percent confidence interval
99 percent confidence interval
Standard error of the proportion

PART III
DECISION MAKING

8 Testing Differences Between Means

IN CHAPTER 7, we saw that a population mean or proportion can be estimated from the information we gain in a single sample. For instance, we might estimate the level of anomie in a certain city, the proportion of aged persons who are economically disadvantaged, or the mean attitude toward racial separation among a population of Americans.

Although the descriptive, fact-gathering approach of estimating means and proportions has obvious importance, it *does not* constitute the primary decision-making goal or activity of social research. Quite to the contrary, most social researchers are preoccupied with the task of *testing hypotheses* about the differences between two or more samples.

When testing differences between samples, social researchers ask such questions as: Do Germans differ from Americans with respect to obedience to authority? Do Protestants or Catholics have a higher rate of suicide? Do political conservatives discipline their children more severely than political liberals? (See Chapter 1.) Note that each research question involves making a *comparison* between two groups: conservatives versus liberals, Protestants versus Catholics, Germans versus Americans.

THE NULL HYPOTHESIS: NO DIFFERENCE BETWEEN MEANS

It has become conventional in statistical analysis to set out testing the *null hypothesis*—the hypothesis that *says* two samples have been drawn from the same population. According to the null hypothesis, any observed difference between samples is regarded as a chance occurrence resulting from sampling error alone. Therefore, an obtained difference between two sample means does not represent a true difference between their population means.

123

In the present context, the null hypothesis can be symbolized as

$$M_1 = M_2$$

where

M_1 = mean of the first population
M_2 = mean of the second population

Let us examine null hypotheses for the research questions that were earlier posed:

1. Germans are no more or less obedient to authority than Americans.
2. Protestants have the same suicide rate as Catholics.
3. Political conservatives and liberals discipline their children to the same extent.

It should be noted that the null hypothesis does not deny the possibility of obtaining differences between *sample* means. On the contrary, it seeks to explain such differences between sample means by attributing them to the operation of sampling error. In accordance with the null hypothesis, for example, if we find that a random *sample* of female dentists earns less money ($\overline{X}$ = $12,000) than a random *sample* of male dentists ($\overline{X}$ = $15,000), we do not, on that basis, conclude that the *population* of female dentists earns less money than the *population* of male dentists. Instead, we treat our obtained sample difference ($15,000 − $12,000 = $3,000) as a product of sampling error—that difference which inevitably results from the process of sampling from a given population. As we shall see later, this aspect of the null hypothesis provides an important link with sampling theory.

THE RESEARCH HYPOTHESIS: A DIFFERENCE BETWEEN MEANS

The null hypothesis is generally (although not necessarily) set up with the hope of nullifying it. This makes sense, for most social researchers seek to establish relationships between variables. That is, they are often more interested in finding differences than in determining that differences do not exist. To illustrate, who would bother studying Catholics and Protestants in the hope that their suicide rates *do not* differ? Differences between groups—whether expected on theoretical or empirical grounds—often provide the rationale on which a study is conducted.

If we reject the null hypothesis, if we find our hypothesis of no difference between means fails to hold, we automatically accept the *research hypothesis* that a true population difference does exist. This is often the hoped-for result in social research. The research hypothesis says that the two samples have been taken from populations having different means. It says that the obtained difference between sample means is too large to be accounted for by sampling error.

The research hypothesis for mean differences is symbolized by

$$M_1 \neq M_2$$

where

M_1 = mean of the first population
M_2 = mean of the second population ($\neq$ is read "does not equal")

The following research hypotheses can be specified for the research questions that were earlier posed:

1. Germans differ from Americans with respect to obedience to authority.
2. Protestants do not have the same suicide rate as Catholics.
3. Political liberals differ from political conservatives with respect to permissive child-rearing methods.

SAMPLING DISTRIBUTION OF MEAN DIFFERENCES

In the preceding chapter, we saw that the 98 means from the 98 samples drawn by our eccentric social researcher could be plotted in the form of a sampling distribution of means. In a similar way, let us now imagine that the same eccentric social researcher takes *not one* but *two* random samples *at a time* from a given population of people. Suppose, for example, that he takes a sample of 500 political liberals *and* a sample of 500 political conservatives. To test the research hypothesis that liberals are more or less permissive parents than conservatives, he then questions all sample members about their child-rearing methods (for example: Do you ever punish your children? Do you ever spank them? If so, how often?). From the responses to such questions, he comes up with a measure of child-rearing permissiveness which can be used to compare the liberal and conservative samples. Scores on this measure range from 1 (not permissive) to 10 (very permissive). As graphically illustrated in Figure 8.1, our eccentric social researcher finds his sample of liberals is more permissive ($\overline{X}$ = 8.0) than his sample of conservatives ($\overline{X}$ = 3.0).

We might ask: In light of sampling error, can we expect a difference between 8.0 and 3.0 (8.0 − 3.0 = +5.0) strictly on the basis of chance and chance alone? Must we accept the null hypothesis of no population difference? Or is the obtained sample difference +5.0 large enough to indicate a true population difference between conservatives and liberals with respect to their child-rearing practices?

We were introduced in Chapter 2 to frequency distributions of raw scores from a given population. In Chapter 7, we saw it was possible to construct a sampling distribution of mean scores, a fre-

FIGURE 8.1
The mean difference in permissiveness between samples of liberals and conservatives as taken from a hypothetical population

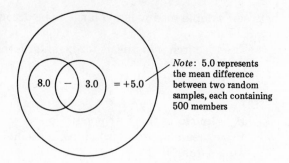

Note: 5.0 represents the mean difference between two random samples, each containing 500 members

quency distribution of sample means. In addressing ourselves to the question at hand, we must now take the notion of frequency distribution a step further and examine the nature of a *sampling distribution of differences,* that is, a frequency distribution of a large number of *differences* between random sample means that have been drawn from a given population.

To illustrate the sampling distribution of differences, let us return to the work of our eccentric social researcher whose passion for drawing random samples has once again led him to continue the sampling process beyond its ordinary limits. Rather than draw a single sample of 500 liberals and a single sample of 500 conservatives, he takes 70 *pairs* of such samples (70 samples *each* containing 500 conservatives, and 70 samples *each* containing 500 liberals). That is, every time he draws 500 conservatives at random, he also draws 500 liberals at random.

Upon taking his samples, our eccentric social researcher questions each and every sample member (1,000 × 70 = 70,000 people) regarding his child-rearing methods and comes up with a mean score of permissiveness for each of the liberal and conservative samples. In addition, he derives a mean difference score by subtracting the conservative mean score from the liberal mean score for each pair of samples. For example, if the mean score of permissiveness for liberals is 7.0, and the mean score of permissiveness for conservatives is 6.0, then the difference score would be + 1.0; likewise, if the liberal mean score is 5.0, and the conservative mean score is 8.0, the difference score would be − 3.0. Obviously, the larger the difference score, the more the two samples differ with respect to the characteristic under investigation. Note that we always subtract the second sample mean from the first sample mean (in the present case, we subtract the conservative mean scores from the liberal mean scores). The 70 mean difference scores derived by our eccentric social researcher have been illustrated in Figure 8.2.

Let us suppose we know that the populations of conservatives and liberals actually do not differ at all with respect to permis-

FIGURE 8.2
Seventy mean difference scores representing differences in permissiveness between liberal and conservative samples taken at random from a hypothetical population

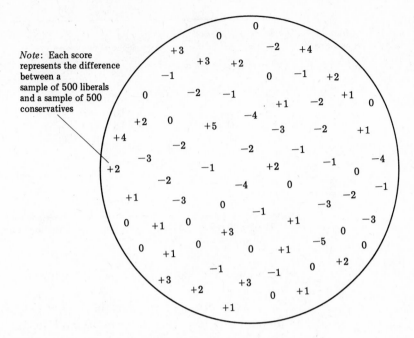

Note: Each score represents the difference between a sample of 500 liberals and a sample of 500 conservatives

siveness in child-rearing methods. Let us say $M = 5.0$ in both liberal and conservative populations. If we assume the null hypothesis is correct, and liberals and conservatives are identical in this respect, we can use the 70 mean differences obtained by our eccentric social researcher to illustrate the sampling distribution of differences. This is true because the sampling distribution of differences makes the assumption that all sample pairs differ only by virtue of sampling error and not as a function of true population differences.

The 70 mean differences in Figure 8.2 have been rearranged as a sampling distribution of mean differences in Table 8.1. Like the scores in other types of frequency distributions, these have been arranged in consecutive order from high to low, while frequency of occurrence is indicated in an adjacent column.

To better depict the key properties of a sampling distribution of differences, the data from Table 8.1 have been graphically presented in Figure 8.3. As illustrated therein, we see that the *sampling distribution of differences between sample means approximates a normal curve whose mean ("mean of differences") is zero.*[1] This makes sense because the positive and negative mean differences in the distribution tend to cancel out one another (for every negative

---

[1] This assumes we have drawn large random samples from a given population of raw scores.

TABLE 8.1
Sampling distribution of differences for 70 pairs of random samples

Mean Difference[a]	f
+5	1
+4	2
+3	5
+2	7
+1	10
0	18
−1	10
−2	8
−3	5
−4	3
−5	1
	N = $\overline{70}$

[a] These difference scores include fractional values (for example, −5 includes the values −5.0 through −5.9).

value, there tends to be a positive value of equal distance from the mean).

As a normal curve, most of the sample mean differences in this distribution fall close to zero—its middlemost point; there are relatively few mean differences having extreme values in either direction from the mean of differences. This is to be expected, since the entire distribution of differences is a product of sampling error rather than actual population differences between conservatives and liberals. In other words, if the actual mean difference between the populations of conservatives and liberals is zero, we also expect the mean of the sampling distribution of differences to be zero.

TESTING HYPOTHESES WITH THE DISTRIBUTION OF DIFFERENCES

In earlier chapters, we learned to make probability statements regarding the occurrence of both raw scores and sample means. In the present case, we seek to make statements of probability about the difference scores in the sampling distribution of mean differences. As pointed out earlier, this sampling distribution takes the

FIGURE 8.3
Frequency polygon of the sampling distribution of differences from Table 8.1

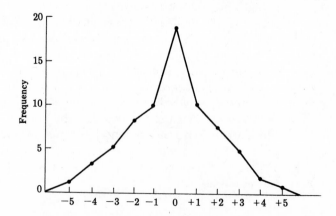

FIGURE 8.4
The sampling distribution of
differences as a probability
distribution

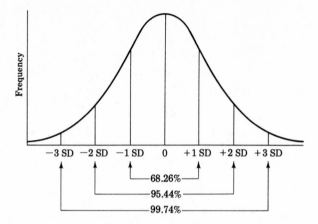

form of the normal curve and, therefore, can be regarded as a probability distribution. We can say that probability decreases as we move farther and farther from the mean of differences (zero). More specifically, as illustrated in Figure 8.4, we see that 68.26 percent of the mean differences fall between −1 SD and +1 SD from zero. In probability terms, this indicates $P = .68$ that any difference between sample means falls within this interval. Similarly, we can say the probability is roughly .95 (95 chances in 100) that any sample mean difference falls between −2 SD and +2 SD from a mean difference of zero, and so on.

The sampling distribution of differences provides a sound basis for testing hypotheses about the mean difference between two random samples. Suppose, for instance, that a sample of 100 liberals has a mean score of permissiveness of 7, while a sample of 100 conservatives has a mean score of permissiveness of 2. The reasoning goes like this: if our obtained mean difference of 5 $(7 - 2 = 5)$ lies so far from a difference of zero that it has only a small *probability* of occurrence in the sampling distribution of differences, we reject the null hypothesis—the hypothesis that says the obtained difference is a result of sampling error. If, on the other hand, our sample mean difference falls so close to zero that its *probability* of occurrence is large, we must accept the null hypothesis and treat our obtained difference as a result of sampling error.

Therefore, we seek to determine how far our obtained mean difference (in this case, 5) lies from a mean difference of zero. In so doing, we must first translate our obtained difference into units of standard deviation.

Recall that we translate *raw scores* into units of standard deviation by the formula

$$z = \frac{X - \overline{X}}{\sigma}$$

where

X = a raw score
$\overline{X}$ = mean of the distribution of raw scores
σ = standard deviation of the distribution of raw scores

Likewise, we translate the *mean scores* in a distribution of sample means into units of standard deviation by the formula

$$z = \frac{\overline{X} - M}{\sigma_{\overline{X}}}$$

where

$\overline{X}$ = a sample mean
M = the population mean (mean of means)
$\sigma_{\overline{X}}$ = standard error of the mean (estimate of standard deviation of the distribution of means)

In the present context, we similarly seek to translate our sample mean difference (+5) into units of standard deviation by the formula

$$z = \frac{(\overline{X}_1 - \overline{X}_2) - 0}{\sigma_{\text{diff}}}$$

where

$\overline{X}_1$ = mean of the first sample
$\overline{X}_2$ = mean of the second sample
"0" = zero, the value of the mean of the sampling distribution of differences (we assume that $M_1 - M_2 = 0$)
σ_{diff} = standard deviation of the sampling distribution of differences

Since the value of the mean of the distribution of differences is always assumed to be zero, we can drop it from the z-score formula without altering our result. Therefore,

$$z = \frac{\overline{X}_1 - \overline{X}_2}{\sigma_{\text{diff}}}$$

With regard to permissiveness among liberals and conservatives, we must first translate our obtained mean difference into its z-score equivalent. If the standard deviation of the sampling distribution of differences (σ_{diff}) is 2, we obtain the following z score:

$$z = \frac{7 - 2}{2}$$

$$= \frac{5}{2}$$

$$= +2.5$$

Thus, a mean difference of 5 between liberals and conservatives falls 2.5 standard deviations from a mean difference of zero in the distribution of differences.

We ask: *What is the probability that a difference of 5 or more between sample means can happen strictly on the basis of sampling error?* Going to Table B in the back of the text, we learn that $z = 2.5$ represents 49.38 percent of the distribution in *either direction* from the mean of zero. That is, 98.76 percent (49.38 percent + 49.38 percent = 98.76 percent) of the sample mean differences lie between zero and a mean difference of 5 in *both directions* from zero, plus *and* minus (see Figure 8.5). In probability terms, this indicates $P = .99$ (99 chances out of 100) that a mean difference will fall between -5 and $+5$. Subtracting from 100 percent (100 percent − 98.76 percent = 1.24 percent), we find $P = .01$ (rounded off) that a mean difference of 5 (or greater than 5) between samples can happen strictly on the basis of sampling error. That is, a mean difference of 5 or more occurs by sampling error (and therefore appears in the sampling distribution) *only once* in every 100 mean differences. Knowing this, would we not consider rejecting the null hypothesis and accepting the research hypothesis that a population difference actually exists between conservatives and liberals with respect to child-rearing permissiveness? One chance out of 100 represents pretty good odds, does it not?

Given the situation above, most of us would choose to reject the null hypothesis, even though we might be wrong in doing so (don't forget that 1 chance out of 100 still remains). However, the decision is not always so clear-cut. Suppose, for example, we learn our mean difference happens by sampling error 10 ($P = .10$), 15 ($P = .15$), or 20 ($P = .20$) times out of 100. Do we still reject the null hypothesis? Or do we "play it safe" and attribute our obtained difference to sampling error?

We need a consistent cutoff point for deciding whether a difference between two sample means is so large that it can no longer be attributed to sampling error. We need a method for determining when our result is *statistically significant*.

FIGURE 8.5
Graphic representation of the percent of total area in the distribution of differences between z = −2.5 and z = +2.5

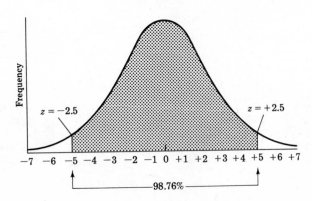

**LEVELS OF
CONFIDENCE**

To establish whether our obtained sample difference is statistically significant—the result of a real population difference and not just sampling error—it is customary to set up a *confidence level* (also known as a *significance level*), a level of probability at which the null hypothesis can be rejected with confidence, and the research hypothesis can be accepted with confidence. Accordingly, we decide to reject the null hypothesis if the probability is very small (for example, only 5 chances out of 100) that the sample difference is a product of sampling error.

It is a matter of convention to use the *.05 level of confidence*. That is, we are willing to reject the null hypothesis, if an obtained sample difference occurs by chance only 5 times or less out of 100 (5 percent). The .05 confidence level has been graphically depicted in Figure 8.6. As shown therein, the .05 level of confidence is found in the small areas of the "tails" of the distribution of mean differences. These are the areas under the curve that represent a distance of plus or minus 1.96 standard deviations from a mean difference of zero.

To better understand why this particular point in the sampling distribution represents the .05 level of confidence, we might turn to Table B in the back of the text to determine the percent of total frequency associated with 1.96 standard deviations from the mean. We see that 1.96 standard deviations in *either* direction represent 2.5 percent of the sample mean differences (50 percent − 47.5 percent = 2.5 percent). In other words, 95 percent of the sample differences fall between − 1.96 SD and + 1.96 SD from a mean difference of zero; only 5 percent fall at or beyond this point (2.5 percent + 2.5 percent = 5 percent).

Confidence levels can be set up for any degree of probability. For instance, a more stringent confidence level is the *.01 level of confidence,* whereby the null hypothesis is rejected if there is only 1 chance out of 100 that the obtained sample difference could occur by sampling error (1 percent). The .01 level of confidence is represented by the area that lies 2.58 standard deviations in both directions from a mean difference of zero.

FIGURE 8.6
*A graphic representation of
the .05 level of confidence*

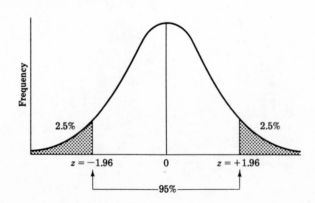

Levels of confidence do not give us an *absolute* statement as to the correctness of the null hypothesis. Whenever we decide to reject the null hypothesis at a certain level of confidence, we open ourselves to the chance of making the wrong decision. Rejecting the null hypothesis when we should have accepted it is known as *alpha error (or type I error)*. The probability of making alpha error can only arise when we reject the null hypothesis and varies according to the level of confidence we choose. For example, if we reject the null hypothesis at the .05 level of confidence and conclude that conservatives actually differ from liberals in terms of child-rearing methods, then there are 5 chances out of 100 we are wrong. In other words, $P = .05$ that we have committed alpha error, and conservatives really do not differ from liberals. Likewise, if we choose the .01 level of confidence, there is only 1 chance out of 100 ($P = .01$) to make the wrong decision regarding the difference between liberals and conservatives. Obviously, the more stringent our level of confidence (the farther out in the tail it lies), the less likely we are to make alpha error. To take an extreme example, setting up a .001 confidence level yields a risk that alpha error occurs only one time in every thousand.

The farther out in the tail of the curve our confidence level falls, however, the greater the risk of making another kind of error known as *beta error (or type II error)*. This is the error of accepting the null hypothesis when we should have rejected it. Beta error indicates that our research hypothesis may still be correct, despite the decision to reject it and accept the null hypothesis. One method for reducing the risk of committing beta error is to increase the size of the samples, so that a true population difference is more likely to be represented.

We can never be certain that we have not made a wrong decision with respect to the null hypothesis, for we examine only a sample and not the complete population. So long as we do not have knowledge of true population values, we take the risk of making either a type I or a type II error, depending on our decision. This is the risk of statistical decision making which the social researcher must be willing to take.

STANDARD ERROR OF THE DIFFERENCE

We can never have first-hand knowledge of the standard deviation of the distribution of mean differences. And just as in the case of the sampling distribution of means (Chapter 7), it would be a major effort if we were actually to draw a large number of sample pairs in order to calculate it. Yet this standard deviation plays an important role in the method for testing hypotheses about mean differences and, therefore, cannot be ignored.

Fortunately, we do have a simple method whereby the standard deviation of the distribution of differences can be accurately estimated on the basis of the two samples we have actually drawn. This estimate of the standard deviation of the sampling distribution of differences is referred to as *the standard error of the difference* and is symbolized by σ_{diff}. By formula,

$$\sigma_{\text{diff}} = \sqrt{\sigma_{\bar{X}_1}^2 + \sigma_{\bar{X}_2}^2}$$

where

σ_{diff} = standard error of the difference
$\sigma_{\bar{X}_1}$ = standard error of the first sample mean
$\sigma_{\bar{X}_2}$ = standard error of the second sample mean

For purposes of illustration, suppose we have obtained the following data for a sample of 50 liberals and a sample of 50 conservatives:

Liberals ($N = 50$)	Conservatives ($N = 50$)
$\bar{X} = 7.0$	$\bar{X} = 6.0$
$s = 2.0$	$s = 1.5$

In order to compute the standard error of the difference, we must first find the standard error for each sample mean. Recall that this is done as follows from the standard deviation for each sample (see Chapter 7):

$$\sigma_{\bar{X}_1} = \frac{s_1}{\sqrt{N_1 - 1}} \qquad \sigma_{\bar{X}_2} = \frac{s_2}{\sqrt{N_2 - 1}}$$

$$= \frac{2.0}{\sqrt{50 - 1}} \qquad = \frac{1.5}{\sqrt{50 - 1}}$$

$$= \frac{2.0}{7.0} \qquad = \frac{1.5}{7.0}$$

$$= .29 \qquad = .21$$

Once we know $\sigma_{\bar{X}}$ for each sample mean, we can obtain σ_{diff} as follows:

$$\sigma_{\text{diff}} = \sqrt{\sigma_{\bar{X}_1}^2 + \sigma_{\bar{X}_2}^2}$$
$$= \sqrt{.29^2 + .21^2}$$
$$= \sqrt{.08 + .04}$$
$$= \sqrt{.12}$$
$$= .35$$

The standard error of the difference (our estimate of the standard deviation of the distribution of differences) turns out to be .35. If we were testing the difference between liberals ($\bar{X} = 7.0$) and conservatives ($\bar{X} = 6.0$) with respect to permissiveness, we could use our result to translate the obtained sample mean difference into its z-score equivalent:

$$z = \frac{\overline{X}_1 - \overline{X}_2}{\sigma_{\text{diff}}}$$

$$= \frac{7 - 6}{.35}$$

$$= \frac{1}{.35}$$

$$= 2.86$$

Turning to Table B in the back of the book, we see that a z score of 2.86 represents exactly 49.79 percent of the mean differences on *either* side or 99.58 percent of the mean differences on *both* sides of a mean difference of zero (49.79 percent + 49.79 percent = 99.58 percent). If we subtract this sum from 100 percent, we learn that less than 1 percent (.42 percent) of the mean difference scores have a value of 1 or larger than 1. Therefore, P is less than .01 to obtain a mean difference of 1 on the basis of sampling error. We can reject the null hypothesis at either the .05 level of confidence or the .01 level of confidence, whichever we have established for our study.

An Illustration

To provide a step-by-step illustration of the foregoing procedure for testing a difference between two sample means, let us say we wanted to test the null hypothesis at the .05 confidence level that females are no more or less ethnocentric than males ($M_1 = M_2$). Our research hypothesis is that females differ from males with respect to ethnocentrism[2] ($M_1 \neq M_2$). To test this hypothesis, let us say we gave a measure of ethnocentrism (for example, the Ethnocentrism Scale) to a random sample of 35 females and a random sample of 35 males, and obtained the following scores of ethnocentrism for each sample (X = scores ranging from 1 representing low ethnocentrism to 5 representing high ethnocentrism):

Males ($N = 35$)		Females ($N = 35$)	
X_1	X^2	X_2	X^2
1	1	1	1
1	1	1	1
1	1	2	4
2	4	1	1
1	1	1	1
1	1	1	1
3	9	3	9
3	9	1	1

---

[2] "Ethnocentrism" refers to the tendency to evaluate all groups of persons using our own cultural standards.

(Continued)

Males (N = 35)		Females (N = 35)	
X_1	X^2	X_2	X^2
1	1	2	4
2	4	4	16
1	1	1	1
2	4	1	1
1	1	1	1
1	1	1	1
1	1	5	25
1	1	1	1
2	4	2	4
4	16	2	4
5	25	1	1
1	1	1	1
1	1	1	1
2	4	1	1
1	1	2	4
2	4	3	9
1	1	1	1
2	4	1	1
1	1	1	1
1	1	2	4
1	1	2	4
1	1	2	4
3	9	1	1
3	9	1	1
1	1	1	1
4	16	1	1
$\Sigma X = 60$	$\Sigma X^2 = 142$	$\Sigma X = 54$	$\Sigma X^2 = 114$

STEP 1: Find the Mean for Each Sample

$$\overline{X}_1 = \frac{\Sigma X_1}{N} \qquad \overline{X}_2 = \frac{\Sigma X_2}{N}$$

$$= \frac{60}{35} \qquad\qquad = \frac{54}{35}$$

$$= 1.71 \qquad\qquad = 1.54$$

STEP 2: Find the Standard Deviation for Each Sample

$$s_1 = \sqrt{\frac{\Sigma X^2}{N} - \overline{X}^2} \qquad s_2 = \sqrt{\frac{\Sigma X^2}{N} - \overline{X}^2}$$

$$= \sqrt{\frac{142}{35} - 2.92} \qquad = \sqrt{\frac{114}{35} - 2.37}$$

$$= \sqrt{4.06 - 2.92} \qquad = \sqrt{3.26 - 2.37}$$

$$= \sqrt{1.14} \qquad\qquad = \sqrt{.89}$$

$$= 1.07 \qquad\qquad = .94$$

STEP 3: Find the Standard Error of Each Mean

$$\sigma_{\overline{X}_1} = \frac{s_1}{\sqrt{N-1}} \qquad \sigma_{\overline{X}_2} = \frac{s_2}{\sqrt{N-1}}$$

$$= \frac{1.07}{\sqrt{34}} \qquad = \frac{.94}{\sqrt{34}}$$

$$= \frac{1.07}{5.83} \qquad = \frac{.94}{5.83}$$

$$= .18 \qquad\qquad = .16$$

STEP 4: Find the Standard Error of the Difference

$$\sigma_{\text{diff}} = \sqrt{\sigma_{\overline{X}_1}{}^2 + \sigma_{\overline{X}_2}{}^2}$$
$$= \sqrt{(.18)^2 + (.16)^2}$$
$$= \sqrt{.03 + .03}$$
$$= \sqrt{.06}$$
$$= .25$$

STEP 5: Translate the Sample Mean Difference into Units of Standard Error of the Difference

$$z = \frac{\overline{X}_1 - \overline{X}_2}{\sigma_{\text{diff}}}$$
$$= \frac{1.71 - 1.54}{.25}$$
$$= \frac{.17}{.25}$$
$$= .68$$

STEP 6: Find the Percent of Total Area Under the Normal Curve Between z and a Mean Difference of Zero (See Table B)

$$\begin{array}{r} 25.17\% \\ + 25.17\% \\ \hline 50.34\% \end{array}$$

STEP 7: Subtract from 100 Percent to Find the Percent of Total Area Associated with the Obtained Sample Mean Difference

$$\begin{array}{r} 100.00\% \\ - 50.34\% \\ \hline 49.66\% \end{array}$$

From the result of Step 7 above, we see $P = .50$ (rounded off) of obtaining a mean difference of .17 (1.71 − 1.54) by sampling error. As a result, we must accept the null hypothesis and reject the research hypothesis at the .05 level of confidence. The probability of occurrence of our obtained mean difference between males and females is *greater* than 5 out of 100. To be exact, it is 50 out of 100! Conclusion: Our sample data do not indicate that females are any more or less ethnocentric than males.

MAKING COMPARISONS BETWEEN SMALL SAMPLES

Social researchers often work with samples that contain a small number of respondents or cases (for example, under 30). While results based on small sample sizes might be convenient, if not necessary, to obtain, they can be seriously misleading if interpreted by the area under the normal curve in Table B. This is true because the sampling distribution of differences takes the form of the normal curve only if the samples that go to make it up are large. A social researcher working with 5, 10, or 20 respondents in each sample cannot meet this assumption. As a result, he cannot use z scores based on the normal distribution.

To statistically compensate for this departure from normality in the distribution of differences, we obtain instead what is commonly known as a t ratio. Like a z score, the t ratio can be used to translate a sample mean difference into units of standard error of the difference. Also like a z score, we obtain a t ratio by taking the difference between our sample means and dividing by our standard error of the difference. By formula,

$$t = \frac{\overline{X}_1 - \overline{X}_2}{\sigma_{\text{diff}}}$$

where

$\overline{X}_1$ = mean of the first sample
$\overline{X}_2$ = mean of the second sample
σ_{diff} = standard error of the difference

As shown above, the t-ratio formula is identical with the formula for a z score which we learned earlier. Unlike a z score, however, the t ratio must be interpreted with reference to *degrees of freedom*[3] (df), which varies directly with sample size and goes to determine the shape of the sampling distribution of differences.

"Degrees of freedom" indicates the extent to which a researcher is free to reject the null hypothesis when it should be rejected; it represents statistical compensation for the failure of the sampling distribution to assume the shape of the normal curve. The larger the sample size, the greater our degrees of freedom. The greater our degrees of freedom, the closer the distribution of differences comes to an approximation of the normal curve. With infinite degrees of freedom, our t ratio becomes a z score, and we may employ Table B to interpret our result.

But what happens when we do work with small samples? How

[3] Degrees of freedom technically refers to freedom of variation among a set of scores. If we have a sample of 6 scores, then 5 are free to vary while only one is fixed in value. Therefore, in a single sample of 6 respondents, df = $N - 1$ or 5.

do we go about the business of finding degrees of freedom and interpreting our t ratio? For a t ratio that represents two sample means, the number of degrees of freedom can be found by the formula

$$df = N_1 + N_2 - 2$$

where

N_1 = size of the first sample
N_2 = size of the second sample

Therefore, if we are comparing a sample of 6 liberals and 8 conservatives, our degrees of freedom would be $6 + 8 - 2 = 12$.

With the aid of Table C in the back of the text and our calculated number of degrees of freedom, we can interpret any t ratio we might obtain. Table C gives the values of t required to reject the null hypothesis at the .05 and .01 levels of confidence for various degrees of freedom. Turning to Table C, we see a column marked df (degrees of freedom) and a list of t values for each degree of freedom at the .05 and .01 confidence levels. As we shall see, these t values can be used to interpret our calculated t ratio.

An Illustration of a Small Sample Comparison

To illustrate the use of the t ratio, degrees of freedom, and Table C for testing a mean difference between small samples, consider the following research situation: A social researcher seeks to test the hypothesis that charitable behavior will vary whether the donation is made anonymously or with the donor's identity made known. Therefore,

Null Hypothesis: *Degree of charitable behavior does not differ ac-*
$(M_1 = M_2)$ *cording to whether a donation is made anony-*
mously or with the donor's identity made known.

Research Hypothesis: *Degree of charitable behavior differs accord-*
$(M_1 \neq M_2)$ *ing to whether a donation is made anony-*
mously or with the donor's identity made
public.

To test this hypothesis, the researcher specifies the .05 level of confidence; that is, he initially chooses to reject the null hypothesis only if it turns out there are 5 chances in 100 that the obtained sample mean difference is a result of sampling error. Having established this criterion of significance, he obtains two random samples of potential donors. From all of the members of both samples, he asks for donations of money to be distributed among the survivors of a major earthquake. To the 6 members of the first sample, he gives his assurance of complete anonymity; to the 6 members of the second sample, he promises to list the names of donors in a conspicuous public place. Hence, we have the experimental conditions of *anonymity* versus *known identity*.

The amounts of money donated by the members of both samples have been listed below:

Anonymity ($N = 6$)		Known Identity ($N = 6$)	
X_1	X^2	X_2	X^2
$1	1	$3	9
2	4	5	25
1	1	5	25
1	1	5	25
2	4	4	16
1	1	5	25
$\Sigma X = 8$	$\Sigma X^2 = 12$	$\Sigma X = 27$	$\Sigma X^2 = 125$

We see that the 6 members of the anonymity sample gave $8 while the 6 members of the known identity sample gave $27. The following step-by-step procedure can be used to test the statistical significance of the obtained difference.

STEP 1: Find the Mean for Each Sample

$$\overline{X}_1 = \frac{\Sigma X_1}{N} \qquad \overline{X}_2 = \frac{\Sigma X_2}{N}$$

$$= \frac{8}{6} \qquad\qquad = \frac{27}{6}$$

$$= \$1.33 \qquad\quad = \$4.50$$

STEP 2: Find the Standard Deviation for Each Sample

$$s_1 = \sqrt{\frac{\Sigma X^2}{N} - \overline{X}^2} \qquad s_2 = \sqrt{\frac{\Sigma X^2}{N} - \overline{X}^2}$$

$$= \sqrt{\frac{12}{6} - (1.33)^2} \qquad = \sqrt{\frac{125}{6} - (4.50)^2}$$

$$= \sqrt{2.00 - 1.77} \qquad = \sqrt{20.83 - 20.25}$$

$$= \sqrt{.23} \qquad\qquad = \sqrt{.58}$$

$$= .48 \qquad\qquad\quad = .76$$

STEP 3: Find the Standard Error of Each Mean

$$\sigma_{\overline{X}_1} = \frac{s}{\sqrt{N-1}} \qquad \sigma_{\overline{X}_2} = \frac{s}{\sqrt{N-1}}$$

$$= \frac{.48}{\sqrt{5}} \qquad\qquad = \frac{.76}{\sqrt{5}}$$

$$= \frac{.48}{2.24} \qquad\qquad = \frac{.76}{2.24}$$

$$= .21 \qquad\qquad\quad = .34$$

STEP 4: Find the Standard Error of the Difference

$$\sigma_{\text{diff}} = \sqrt{\sigma_{\overline{X}_1}{}^2 + \sigma_{\overline{X}_2}{}^2}$$
$$= \sqrt{(.21)^2 + (.34)^2}$$
$$= \sqrt{.04 + .12}$$
$$= \sqrt{.16}$$
$$= .40$$

STEP 5:   Translate the Sample Mean Difference into Units of Standard Error of the Difference

$$t = \frac{\overline{X}_1 - \overline{X}_2}{\sigma_{\text{diff}}}$$
$$= \frac{1.33 - 4.50}{.40}$$
$$= -\frac{3.17}{.40}$$
$$= -7.93$$

STEP 6:   Find the Number of Degrees of Freedom

$$df = N_1 + N_2 - 2$$
$$= 6 + 6 - 2$$
$$= 10$$

STEP 7:   Compare the Obtained t Ratio with the Appropriate t Ratio in Table C

$$\text{obtained } t \text{ ratio} = 7.93$$
$$\text{table } t \text{ ratio} = 2.228$$
$$df = 10$$
$$P = .05$$

As shown in Step 7, in order to reject the null hypothesis at the .05 level of confidence with 10 degrees of freedom, our calculated t ratio must be 2.228 or larger. In the present case, we have obtained a t ratio of 7.93. Therefore, we reject the null hypothesis and accept the research hypothesis. The degree of charitable behavior actually varies according to whether a donation is made anonymously or with the donor's identity made known. More specifically, the "identity known" condition produces significantly more charity ($\overline{X} = \$4.50$) than does the "anonymous" condition ($\overline{X} = \$1.33$).

MAKING COMPARISONS BETWEEN SAMPLES OF UNEQUAL SIZE

Until now, we have worked with samples that contain exactly the same number of respondents or cases. In the illustration above, for instance, each sample contained 6 respondents. When we actually go out and conduct research, however, we often find that our samples differ in size. We might have a sample of 50 liberals and a sample of 64 conservatives, a sample of 15 males and 22 females. To make comparisons between samples of unequal size, we must find a way to give appropriate *weight* to the relative influence of each sample. In the case of $\overline{X}$, this is done automatically, for we always divide ΣX by N. Such is not the case for the standard error of the difference: each sample standard deviation on which σ_{diff} is based contributes equally to the formula that we earlier learned, although large and important differences in sample size may be present.

This problem can be overcome by using a formula for the standard error of the difference in which the relative influence of each standard deviation can be weighted in terms of its sample size. Such a formula is presented below:

$$\sigma_{\text{diff}} = \sqrt{\left(\frac{N_1 s_1^2 + N_2 s_2^2}{N_1 + N_2 - 2}\right)\left(\frac{1}{N_1} + \frac{1}{N_2}\right)}$$

where

s_1 = standard deviation of the first sample
s_2 = standard deviation of the second sample
N_1 = total number in the first sample
N_2 = total number in the second sample

To illustrate the step-by-step procedure for comparing samples of unequal size, consider the hypothesis that children with and without learning disabilities in a certain urban neighborhood differ with regard to the tendency toward delinquency. In this case,

Null Hypothesis: *Children with and without learning disabilities do*
$(M_1 = M_2)$ *not differ with regard to the tendency toward delinquency.*

Research Hypothesis: *Children with and without learning disabilities differ with regard to the tendency toward delinquency.*
$(M_1 \neq M_2)$

To test this issue at the .05 level of confidence, imagine that a particular researcher administered a measure of "tendency toward delinquency" to a random sample of 4 children with learning disabilities and a random sample of 7 children without learning disabilities. The following scores of "tendency toward delinquency" resulted (scores range from 1 representing little tendency toward delinquency to 5 representing strong tendency toward delinquency):

Children with Disabilities ($N = 4$)		Children without Disabilities ($N = 7$)	
X_1	X^2	X_2	X^2
1	1	4	16
2	4	1	1
1	1	1	1
3	9	1	1
$\Sigma X_1 = 7$	$\Sigma X^2 = 15$	2	4
		2	4
		1	1
		$\Sigma X_2 = 12$	$\Sigma X^2 = 28$

The step-by-step procedure for testing the foregoing hypothesis can be illustrated as follows:

STEP 1:   Find the Mean for Each Sample

$$\overline{X}_1 = \frac{\Sigma X_1}{N} \qquad \overline{X}_2 = \frac{\Sigma X_2}{N}$$

$$= \frac{7}{4} \qquad\qquad = \frac{12}{7}$$

$$= 1.75 \qquad\quad = 1.71$$

STEP 2:   Find the Standard Deviation for Each Sample

$$s_1 = \sqrt{\frac{\Sigma X^2}{N} - \overline{X}^2} \qquad s_2 = \sqrt{\frac{\Sigma X^2}{N} - \overline{X}^2}$$

$$= \sqrt{\frac{15}{4} - 3.06} \qquad = \sqrt{\frac{28}{7} - 2.92}$$

$$= \sqrt{3.75 - 3.06} \qquad = \sqrt{4.00 - 2.92}$$

$$= \sqrt{.69} \qquad\qquad = \sqrt{1.08}$$

$$= .83 \qquad\qquad\quad = 1.04$$

STEP 3:   Find the Standard Error of the Difference

$$\sigma_{\text{diff}} = \sqrt{\left(\frac{Ns_1{}^2 + Ns_2{}^2}{N_1 + N_2 - 2}\right)\left(\frac{1}{N_1} + \frac{1}{N_2}\right)}$$

$$= \sqrt{\left(\frac{4(.83)^2 + 7(1.04)^2}{4 + 7 - 2}\right)\left(\frac{1}{4} + \frac{1}{7}\right)}$$

$$= \sqrt{\left(\frac{2.76 + 7.56}{9}\right)(.25 + .14)}$$

$$= \sqrt{\left(\frac{10.32}{9}\right)(.39)}$$

$$= \sqrt{(1.15)(.39)}$$

$$= \sqrt{.45}$$

$$= .67$$

STEP 4:   Translate the Sample Mean Difference into Units of Standard Error of the Difference

$$t = \frac{\overline{X}_1 - \overline{X}_2}{\sigma_{\text{diff}}}$$

$$= \frac{1.75 - 1.71}{.67}$$

$$= \frac{.04}{.67}$$

$$= .06$$

STEP 5:   Find the Number of Degrees of Freedom

$$\begin{aligned} df &= N_1 + N_2 - 2 \\ &= 4 + 7 - 2 \\ &= 9 \end{aligned}$$

STEP 6:   Compare the Obtained t Ratio with the Appropriate t Ratio in Table C

$$\begin{aligned} \text{obtained } t \text{ ratio} &= .06 \\ \text{table } t \text{ ratio} &= 2.262 \\ df &= 9 \\ P &= .05 \end{aligned}$$

As indicated in Step 6, to reject the null hypothesis at the .05 level of confidence with 9 degrees of freedom, our obtained t ratio would have to be 2.262 or larger. Since we have calculated a t ratio of only .06, we must accept the null hypothesis and reject the research hypothesis. Our results do not support the view that children with and without learning disabilities differ with regard to the tendency toward delinquency.

COMPARING THE SAME SAMPLE MEASURED TWICE

So far, we have discussed making comparisons between two independently drawn samples (for example, males versus females, blacks versus whites, or liberals versus conservatives). Before leaving this topic, we must now introduce a final variation of the two mean comparison referred to as a *before-after* or *panel* design: the case of a *single* sample measured at two different points in time (time 1 versus time 2). For example, a survey researcher may seek to measure hostility in a single sample of children both before and after they watch a certain television program. In the same way, we might want to measure differences in attitudes toward a particular candidate for office before and after a campaign advertisement appears.

To provide a step-by-step illustration of a before-after comparison, let us suppose that several individuals have been forced by a city government to relocate their homes in order to make way for highway construction. As social researchers, we are interested in determining the impact of forced residential mobility on feelings of neighborliness (that is, positive feelings about neighbors in the *pre*relocation neighborhood versus feelings about neighbors in the *post*relocation neighborhood). In this case, then, M_1 is the mean score of neighborliness at time 1 (*before* relocating), and M_2 is the mean score of neighborliness at time 2 (*after* relocating). Therefore,

Null Hypothesis: *Degree of neighborliness does not differ before and* ($M_1 = M_2$)   *after the relocation.*

Research Hypothesis: *Degree of neighborliness differs before and* ($M_1 \neq M_2$)   *after the relocation.*

To test the impact of forced relocation on neighborliness, we interview a random sample of 6 individuals about their neighbors both before and after they are forced to move. Our interviews yield the following scores of neighborliness (higher scores from 1 to 4 indicate greater neighborliness):

Respondent	Before Move X_1	After Move X_2	Difference $X_1 - X_2 = D$	(Difference)² D^2
Stephanie	2	1	1	1
Myron	1	2	−1	1
Carol	3	1	2	4
Linda	3	1	2	4
Allan	1	2	−1	1
David	4	1	3	9
	$\Sigma X_1 = 14$	$\Sigma X_2 = 8$		$\Sigma D^2 = 20$

As shown above, making a before-after comparison focuses our attention on the *difference* between time 1 and time 2, as reflected in the formula to obtain the standard deviation (for the distribution of before/after difference scores):

$$s = \sqrt{\frac{\Sigma D^2}{N} - (\overline{X}_1 - \overline{X}_2)^2}$$

where

s = standard deviation of the distribution of before/after difference scores

D = "after" raw score subtracted from "before" raw score

N = number of cases or respondents in sample

STEP 1: Find the Mean for Each Point in Time

$$\overline{X}_1 = \frac{\Sigma X_1}{N} \qquad \overline{X}_2 = \frac{\Sigma X_2}{N}$$

$$= \frac{14}{6} \qquad\qquad = \frac{8}{6}$$

$$= 2.33 \qquad\qquad = 1.33$$

STEP 2: Find the Standard Deviation for the Difference Between Time 1 and Time 2

$$s = \sqrt{\frac{\Sigma D^2}{N} - (\overline{X}_1 - \overline{X}_2)^2}$$

$$= \sqrt{\frac{20}{6} - (2.33 - 1.33)^2}$$

$$= \sqrt{\frac{20}{6} - 1.00}$$

$$= \sqrt{3.33 - 1.00}$$

$$= \sqrt{2.33}$$

$$= 1.53$$

STEP 3:   Find the Standard Error of the Difference

$$\sigma_{\text{diff}} = \frac{s}{\sqrt{N-1}}$$

$$= \frac{1.53}{\sqrt{6-1}}$$

$$= \frac{1.53}{2.24}$$

$$= .68$$

STEP 4:   Translate the Sample Mean Difference into Units of Standard Error of the Difference

$$t = \frac{\overline{X}_1 - \overline{X}_2}{\sigma_{\text{diff}}}$$

$$= \frac{2.33 - 1.33}{.68}$$

$$= \frac{1.00}{.68}$$

$$= 1.47$$

STEP 5:   Find the Number of Degrees of Freedom

$$\text{df} = N - 1 \qquad \textit{Note: } N \text{ refers to the total number of } \textit{cases,} \text{ not}$$
$$= 6 - 1 \qquad\qquad \text{the number of scores for which there are}$$
$$= 5 \qquad\qquad\quad 2 \text{ per case or respondent.}$$

STEP 6:   Compare the Obtained t Ratio with the Appropriate t Ratio in Table C

$$\text{obtained } t \text{ ratio} = 1.47$$
$$\text{table } t \text{ ratio} = 2.571$$
$$\text{df} = 5$$
$$P = .05$$

In order to reject the null hypothesis at the .05 confidence level with 5 degrees of freedom, we must obtain a calculated t ratio of 2.571. Since our t ratio is only 1.47—less than the required table value—we accept the null hypothesis and reject the research hypothesis. The obtained sample difference in neighborliness before and after the relocation was actually a result of sampling error.

REQUIREMENTS FOR THE USE OF THE z SCORE AND t RATIO

As we shall see throughout the remainder of this text, every statistical test should be used only if the social researcher has, at least, considered certain requirements, conditions, or assumptions. Employing a test inappropriately may confuse an issue and mislead the investigator. As a result, the following requirements should be kept firmly in mind when considering the appropriateness of the z score or t ratio as a test of significance.

1. A comparison between two means—the z score and t ratio are employed in order to make comparisons between two means from independent samples or from a single sample arranged in a before-after panel design.
2. Interval data—the assumption is that we have scores at the interval level of measurement. Therefore, we cannot use the z score or t ratio for ranked data or data that can only be categorized at the nominal level of measurement (see Chapter 1).
3. Random sampling—we should have drawn our samples on a random basis from a population of scores.
4. A normal distribution—the t ratio for small samples requires that the sample characteristic we have measured be normally distributed in the underlying population (the z score for large samples is not much affected by failure to meet this assumption). Often, we cannot be 100 percent certain that normality exists. Having no reason to believe otherwise, many researchers pragmatically assume their sample characteristic to be normally distributed. However, if the researcher has reason to suspect that normality cannot be assumed, he or she is best advised that the t ratio may be an inappropriate test (see Chapter 6).

SUMMARY

This chapter has focused on testing hypotheses about differences between sample means. Relating to this purpose, the sampling distribution of mean differences as a probability distribution was described and illustrated. With the aid of this distribution and the standard error of the difference, a probability statement about a mean difference could be made and, on that basis, a null hypothesis rejected or accepted at a specified level of confidence. In addition, we saw that a t ratio (and degrees of freedom) could be used to test hypotheses about differences between small samples, between samples of unequal size, and for a single sample measured at two points in time. The appropriateness of the t ratio depends on certain requirements such as (1) making a comparison between two means, (2) interval data, (3) random sampling, and (4) a normal distribution.

PROBLEMS

1. Two groups of subjects participated in an experiment designed to test the effect of frustration on aggression. The experimental group of 40 subjects received a frustrating puzzle to solve,

whereas the control group of 40 subjects received an easy, non-frustrating version of the same puzzle. Level of aggression was then measured for both groups. While the experimental group (frustration) had a mean aggression score of 4.0 and a standard deviation of 2.0, the control group (no frustration) had a mean aggression score of 3.0 and a standard deviation of 1.5 (higher mean scores indicate greater aggression). Using the data above, test the null hypothesis of no difference with respect to aggression between the frustration and no frustration conditions. What do your results indicate?

2. Two groups of students took final exams in statistics. Only one group was given formal course preparation for the exam; the other group read the required text but never attended lectures. The first group (attendance at lectures) got exam grades of 2, 2, 3, and 4; the second group (no attendance at lectures) received exam grades of 1, 1, 2, and 3. Test the null hypothesis of no difference with respect to exam scores between students in attendance at lectures and students not in attendance at lectures. What do your results indicate? (*Note:* Exams were graded from 1 to 10, with higher scores representing better knowledge of statistics.)

3. Test for the significance of the difference between means of the following random samples of scores:

Sample 1	Sample 2
8	1
3	5
1	8
7	3
7	2
6	1
8	2

4. Test for the significance of the difference between means of the following random samples of scores:

Sample 1	Sample 2
6	6
6	5
8	7
7	7
5	3
4	3
8	5
7	6
7	3

5. Test for the significance of the difference between means of the following random samples of scores:

Sample 1	Sample 2
15	10
18	11
12	12
17	10
19	10

6. Test for the significance of the difference between means of the following random samples of scores:

Sample 1	Sample 2
1	2
1	2
2	4
3	2
3	2

7. Test for the signifiance of the difference between means of the following random samples of scores:

Sample 1	Sample 2
5	10
7	7
7	9
3	9
6	7
5	8
4	
6	
7	

8. Test for the significance of the difference between means of the following random samples of scores:

Sample 1	Sample 2
3	7
6	8
4	8
2	9
1	9
	6
	5

9. Test for the significance of the difference between means of the following random samples of scores:

Sample 1	Sample 2
10	10
4	10
1	8
2	7
4	
8	
3	
5	

10. Both before and after seeing a movie designed to reduce prejudice against minority groups, six students were questioned regarding their attitudes toward Jews. On the data below, test the hypothesis that there was no difference in attitudes toward Jews among these students before and after seeing the movie (higher scores indicate more favorable attitudes toward Jews):

Student	Before	After
A	2	4
B	2	5
C	4	3
D	6	8
E	7	9
F	5	8

11. Test for the significance of the "before-after" difference between means in the following random sample of scores:

Respondent	Before	After
A	7	3
B	6	4
C	5	2
D	4	3

12. Test for the significance of the "before-after" difference between means in the following random sample of scores:

Respondent	Before	After
A	6	3
B	7	4
C	10	9
D	9	7
E	8	5

9 Analysis of Variance

BLACKS VERSUS WHITES, males versus females, and liberals versus conservatives represent the kind of two sample comparisons that occupied our attention in the previous chapter. Yet, social reality cannot always be conveniently sliced into two groups; respondents do not always divide themselves in so simple a fashion.

As a result, the social researcher often seeks to make comparisons among three, four, five, or more samples or groups. To illustrate, he or she may study the influence of racial identity (black, white, or oriental) on job discrimination, degree of economic deprivation (severe, moderate, or mild) on juvenile delinquency, subjective social class (upper, middle, working, or lower) on achievement motivation.

The student may wonder whether we use a *series* of t ratios to make comparisons among three or more sample means. Suppose, for example, we want to test the influence of social class on achievement motivation. Why can we not pair-compare all possible combinations of social class and obtain a t ratio for each comparison? Using this method, four samples generate six paired combinations for which six t ratios must be calculated:

1. upper-class versus middle-class;
2. upper-class versus working-class;
3. upper-class versus lower-class;
4. middle-class versus working-class;
5. middle-class versus lower-class;
6. working-class versus lower-class.

Not only does the procedure of calculating a series of t ratios involve a good deal of work, but it has a statistical limitation as well. This is because it increases the probability of making alpha error—the error of rejecting the null hypothesis when it should be accepted. Recall that the social researcher is generally willing to accept a 5 percent risk of making alpha error (the

.05 level of confidence). He therefore expects that *by chance alone* 5 out of every 100 sample mean differences will be large enough to be considered as significant. The more statistical tests we conduct, however, the more likely we are to get statistically significant findings by sampling error (rather than by a true population difference) and hence to commit alpha error. When we run a large number of such tests, the interpretation of our result becomes problematic. To take an extreme example: how would we interpret a significant t ratio out of 1000 such comparisons made in a particular study? We know that at least a few large mean differences can be expected to occur simply on the basis of sampling error.

To overcome this problem and clarify the interpretation of our result, we need a statistical test that holds alpha error at a constant level by making a *single* overall decision as to whether a significant difference is present among the three or more sample means we seek to compare. Such a test is known as the *analysis of variance.*

THE LOGIC OF ANALYSIS OF VARIANCE

To conduct an analysis of variance, we treat the total *variation* in a set of scores as being divisible into two components: the distance of raw scores from their group mean known as *variation within groups* and the distance of group means from one another referred to as *variation between groups.*

In order to examine variation within groups, the achievement-motivation scores of members of four social classes—(1) lower, (2) working, (3) middle, and (4) upper—have been graphically represented in Figure 9.1, where X_1, X_2, X_3, and X_4 are any raw scores in their respective group, and $\overline{X}_1$, $\overline{X}_2$, $\overline{X}_3$, and $\overline{X}_4$ are the group means. In symbolic terms, we see that variation within groups refers to the distance between X_1 and $\overline{X}_1$, between X_2 and $\overline{X}_2$, between X_3 and $\overline{X}_3$, and between X_4 and $\overline{X}_4$.

We can also visualize variation between groups. With the aid of Figure 9.2, we see that degree of achievement motivation varies by social class: the upper-class group ($\overline{X}_4$) has greater achievement motivation than the middle-class group ($\overline{X}_3$) which, in turn, has greater achievement motivation than the working class group ($\overline{X}_2$) which, in its turn, has greater achievement motivation than the lower-class group ($\overline{X}_1$).

The distinction between variation *within* groups and variation *between* groups is not peculiar to the analysis of variance. Although not named as such, we encountered a similar distinction in the form of the t ratio, wherein a difference *between* $\overline{X}_1$ and $\overline{X}_2$ was compared against the standard error of the difference (σ_{diff}), a combined estimate of differences *within* each group. Therefore,

FIGURE 9.1
Graphic representation of variation within four groups of social class

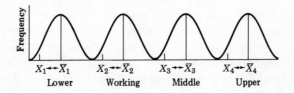

FIGURE 9.2
*Graphic representation of
variation between four
groups of social class*

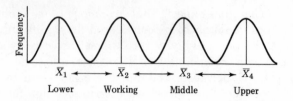

$$t = \frac{\overline{X}_1 - \overline{X}_2}{\sigma_{\text{diff}}} \quad \begin{array}{l} \longleftarrow \text{variation between groups} \\ \longleftarrow \text{variation within groups} \end{array}$$

In a similar way, the analysis of variance yields an *F ratio,* whose numerator represents variation between the groups being compared, and whose denominator contains an estimate of variation within these groups. As we shall see, the *F* ratio indicates the size of the difference between groups *relative* to the size of the variation within each group. As was true of the *t* ratio, the larger the *F* ratio (the larger the variation between groups relative to the variation within groups), the greater the probability of rejecting the null hypothesis and accepting the research hypothesis.

THE SUMS OF SQUARES

At the heart of the analysis of variance is the concept of *sum of squares,* which represents the initial step for measuring total variation as well as variation between and within groups. It may come as a pleasant surprise to learn that only the label "sum of squares" is new to us. The concept itself was introduced in Chapter 5 as an important step in the procedure for obtaining the standard deviation. In that context, we learned to find the sum of squares by squaring the deviations from the mean of a distribution and adding these deviation scores together (Σx^2). This procedure eliminated minus signs, while still providing a sound mathematical basis for the standard deviation.

When applied to a situation in which groups are being compared, there is more than one type of sum of squares, although each type represents *the sum of squared deviations from a mean.* Corresponding to the distinction between total variation and its two components, we have the total sum of squares (SS_{total}), between-groups sum of squares (SS_{bet}), and within-groups sum of squares (SS_{within}).

A Research Illustration

Let us consider a research situation in which each type of sum of squares might be calculated. Suppose we seek to determine the influence of political orientation on child-rearing methods. In the preceding chapter, we tackled this issue by comparing liberals and conservatives. By contrast, we now want to make comparisons representing *several* points along the political continuum. For example, we might compare the child-rearing permissiveness of conservatives, moderates, liberals, and radicals. In such a case,

Null Hypothesis: *Conservatives, moderates, liberals, and radi-*
$(M_1 = M_2 = M_3 = M_4)$ *cals do not differ with respect to child-rearing*
               permissiveness.

Research Hypothesis: Conservatives, moderates, liberals, and radi-
$(M_1 \neq M_2 \neq M_3 \neq M_4)$ *cals differ with respect to child-rearing per-*
               missiveness.

Let us imagine we have actually interviewed random samples of four conservatives, four moderates, four liberals, and four radicals in order to determine their child-rearing methods. Further imagine we have obtained the scores of permissiveness shown in Table 9.1 (scores range from 1 representing little permissiveness to 5 representing much permissiveness).

Within-Groups Sum of Squares

The *within-groups sum of squares* gives us the *sum of the squared deviations of every raw score from its sample mean.* Therefore, the within-groups sum of squares can be obtained by simply *combining* the sum of squares within each sample. By formula,

$$SS_{within} = \Sigma x_1{}^2 + \Sigma x_2{}^2 + \Sigma x_3{}^2 + \Sigma x_4{}^2$$

where

x = a deviation score $(X - \overline{X})$

Applying the SS_{within} formula to the data in Table 9.1, we see

$$SS_{within} = 1.00 + 2.00 + .74 + 2.74$$
$$= 6.48$$

Between-Groups Sum of Squares

The *between-groups sum of squares* represents the *sum of the squared deviations of every sample mean from the total mean.* Accordingly, we must determine the difference between each sample mean and the total mean $(\overline{X} - \overline{X}_{total})$, square this difference score, multiply by the number of scores in the sample, and add these quantities. The definitional formula for the between-groups sum of squares is

$$SS_{bet} = \Sigma(\overline{X} - \overline{X}_{total})^2 N$$

where

$\overline{X}$ = any sample mean
$\overline{X}_{total}$ = total mean (the mean of all raw scores from all samples combined)
N = the number of scores in any sample
SS_{bet} = between-groups sum of squares

The procedure for finding the between-groups sum of squares for the data in Table 9.1 can be summarized as follows:

TABLE 9.1
Scores of child-rearing permissiveness for samples of conservatives, moderates, liberals, and radicals

Conservatives ($N = 4$)			Moderates ($N = 4$)		
X_1	x	x^2	X_2	x	x^2
1	$-.50$	.25	1	-1	1
2	.50	.25	3	1	1
1	$-.50$	.25	2	0	0
2	.50	.25	2	0	0
$\Sigma X = 6$		$\Sigma x^2 = 1.00$	$\Sigma X = 8$		$\Sigma x^2 = 2.00$

$$\overline{X}_1 = \frac{6}{4} = 1.5 \qquad\qquad \overline{X}_2 = \frac{8}{4} = 2.0$$

Liberals ($N = 4$)			Radicals ($N = 4$)		
X_3	x	x^2	X_4	x	x^2
1	$-.75$	.56	3	1.25	1.56
2	.25	.06	2	.25	.06
2	.25	.06	1	$-.75$	.56
2	.25	.06	1	$-.75$	.56
$\Sigma X = 7$		$\Sigma x^2 = .74$	$\Sigma X = 7$		$\Sigma x^2 = 2.74$

$$\overline{X}_3 = \frac{7}{4} = 1.75 \qquad\qquad \overline{X}_4 = \frac{7}{4} = 1.75$$

$$\overline{X}_{\text{total}} = 1.75$$

$$
\begin{aligned}
\text{SS}_{\text{bet}} &= (1.50 - 1.75)^2 4 + (2.0 - 1.75)^2 4 \\
&\quad + (1.75 - 1.75)^2 4 + (1.75 - 1.75)^2 4 \\
&= (-.25)^2 4 + (.25)^2 4 + (0)4 + (0)4 \\
&= (.06)4 + (.06)4 + (0)4 + (0)4 \\
&= .24 + .24 \\
&= .48
\end{aligned}
$$

Total Sum of Squares

It can be shown that the *total sum of squares, the sum of the squared deviations of every raw score from the total mean of the study,* is equal to a combination of its within- and between-group components. The total sum of squares for the data in Table 9.1 can be found as follows:

$$
\begin{aligned}
\text{SS}_{\text{total}} &= \text{SS}_{\text{bet}} + \text{SS}_{\text{within}} \\
&= .48 + 6.48 \\
&= 6.96
\end{aligned}
$$

The total sum of squares can also be defined in terms of the equation

$$\text{SS}_{\text{total}} = \Sigma(X - \overline{X}_{\text{total}})^2$$

where

X = a raw score in any sample
$\overline{X}_{\text{total}}$ = the total mean (the mean of all raw scores from all samples combined)
SS_{total} = the total sum of squares

Using the above formula, we subtract the total mean ($\overline{X}_{total}$) from each raw score in the study (X), square the deviation scores that result, and add them together.

For the data in Table 9.1,

$$
\begin{aligned}
SS_{total} &= (1 - 1.75)^2 + (2 - 1.75)^2 + (1 - 1.75)^2 + (2 - 1.75)^2 \\
&\quad + (1 - 1.75)^2 + (3 - 1.75)^2 + (2 - 1.75)^2 \\
&\quad + (2 - 1.75)^2 + (1 - 1.75)^2 + (2 - 1.75)^2 \\
&\quad + (2 - 1.75)^2 + (2 - 1.75)^2 + (3 - 1.75)^2 \\
&\quad + (2 - 1.75)^2 + (1 - 1.75)^2 + (1 - 1.75)^2 \\
&= (-.75)^2 + (.25)^2 + (-.75)^2 + (.25)^2 + (-.75)^2 \\
&\quad + (1.25)^2 + (.25)^2 + (.25)^2 + (-.75)^2 + (.25)^2 \\
&\quad + (.25)^2 + (.25)^2 + (1.25)^2 + (25)^2 + (-.75)^2 \\
&\quad + (-.75)^2 \\
&= .56 + .06 + .56 + .06 + .56 + 1.56 + .06 \\
&\quad + .06 + .56 + .06 + .06 + .06 + 1.56 + .06 \\
&\quad + .56 + .56 \\
&= 6.96
\end{aligned}
$$

Computing Sums of Squares

The definitional formulas for within-groups, between-groups, and total sums of squares as presented above are based on the manipulation of deviation scores, a time-consuming and difficult requirement. Fortunately, we may, instead, employ the much simpler computational formulas below to obtain a result in the form of an F ratio, which is identical (minus rounding errors) to that obtained with the lengthier definitional formulas.

The raw scores in Table 9.1 have been set up in Table 9.2, for the purpose of illustrating the use of the sum-of-squares computational formulas.

The computational formula for the total sum of squares is the following:

$$
SS_{total} = \Sigma X^2_{total} - \frac{(\Sigma X_{total})^2}{N_{total}}
$$

where

N_{total} = the total number of scores in all samples combined

Carrying out this formula for the data in Table 9.2,

$$
\begin{aligned}
SS_{total} &= (10 + 18 + 13 + 15) - \frac{(6 + 8 + 7 + 7)^2}{4 + 4 + 4 + 4} \\
&= 56 - \frac{(28)^2}{16} \\
&= 56 - \frac{784}{16} \\
&= 56 - 49 \\
&= 7
\end{aligned}
$$

TABLE 9.2
Scores of child-rearing permissiveness for samples of conservatives, moderates, liberals, and radicals

Conservatives ($N = 4$)		Moderates ($N = 4$)	
X_1	X^2	X_2	X^2
1	1	1	1
2	4	3	9
1	1	2	4
2	4	2	4
$\Sigma X = 6$	$\Sigma X^2 = 10$	$\Sigma X = 8$	$\Sigma X^2 = 18$
$\overline{X} = \dfrac{6}{4} = 1.5$		$\overline{X} = \dfrac{8}{4} = 2.0$	

Liberals ($N = 4$)		Radicals ($N = 4$)	
X_3	X^2	X_4	X^2
1	1	3	9
2	4	2	4
2	4	1	1
2	4	1	1
$\Sigma X = 7$	$\Sigma X^2 = 13$	$\Sigma X = 7$	$\Sigma X^2 = 15$
$\overline{X} = \dfrac{7}{4} = 1.75$		$\overline{X} = \dfrac{7}{4} = 1.75$	
		$\overline{X}_{\text{total}} = 1.75$	

The between-groups sum of squares can be obtained by means of the following computational formula:

$$SS_{bet} = \left[\sum \frac{(\Sigma X)^2}{N} \right] - \frac{(\Sigma X_{\text{total}})^2}{N_{\text{total}}}$$

where

N = the total number of scores in any sample

N_{total} = the total number of scores in all samples combined

For example, in Table 9.2,

$$SS_{bet} = \frac{(6)^2}{4} + \frac{(8)^2}{4} + \frac{(7)^2}{4} + \frac{(7)^2}{4} - \frac{(28)^2}{16}$$

$$= \frac{36}{4} + \frac{64}{4} + \frac{49}{4} + \frac{49}{4} - \frac{784}{16}$$

$$= 9.0 + 16 + 12.25 + 12.25 - 49.0$$

$$= 49.5 - 49.0$$

$$= .50$$

Since the within-groups sum of squares is the most time-consuming to compute, we can take advantage of the fact that the total sum of squares is equal to a combination of its two components. Therefore,

$$SS_{within} = SS_{total} - SS_{bet}$$

In the present case,

$$SS_{within} = 7.00 - .50$$
$$= 6.50$$

The following computational formula for the within-groups sum of squares can serve as a check on computational errors:

$$SS_{within} = \sum \left[(\Sigma X^2) - \frac{(\Sigma X)^2}{N} \right]$$

where

X = a raw score in any sample
N = the total number of scores in any sample

Substituting the data in Table 9.2,

$$SS_{within} = \left[10 - \frac{(6)^2}{4} \right] + \left[18 - \frac{(8)^2}{4} \right]$$
$$+ \left[13 - \frac{(7)^2}{4} \right] + \left[15 - \frac{(7)^2}{4} \right]$$
$$= \left(10 - \frac{36}{4} \right) + \left(18 - \frac{64}{4} \right)$$
$$+ \left(13 - \frac{49}{4} \right) + \left(15 - \frac{49}{4} \right)$$
$$= (10 - 9.0) + (18 - 16.0) + (13 - 12.25)$$
$$+ (15 - 12.25)$$
$$= 1.0 + 2.0 + .75 + 2.75$$
$$= 6.50$$

MEAN SQUARE

As we might expect from a measure of variation, the value of the sums of squares tends to become larger as variation increases. For example, SS = 10.9 probably designates greater variation than SS = 1.3. However, sum of squares also gets larger with increasing sample size, so that N = 200 will yield a larger SS than N = 20. As a result, the sum of squares cannot be regarded as an entirely satisfactory "pure" measure of variation, unless, of course, we can find a way to control for the number of scores involved.

Fortunately, such a method exists in a measure of variation known as the *mean square* (or *variance*), which we obtain by dividing SS_{bet} or SS_{within} by the appropriate degrees of freedom (in Chapter 5, we similarly divided Σx^2 by N as a step toward obtaining the standard deviation). Therefore,

$$MS_{bet} = \frac{SS_{bet}}{df_{bet}}$$

where

MS_{bet} = between-groups mean square
SS_{bet} = between-groups sum of squares
df_{bet} = between-groups degrees of freedom

and

$$MS_{within} = \frac{SS_{within}}{df_{within}}$$

where

MS_{within} = within-groups mean square
SS_{within} = within-groups sum of squares
df_{within} = within-groups degrees of freedom

But we must still obtain the appropriate degrees of freedom. For between-groups mean square,

$$df_{bet} = k - 1$$

where

k = the number of samples

To find within-groups mean square,

$$df_{within} = N_{total} - k$$

where

N_{total} = total number of scores in all samples combined
k = the number of samples

Illustrating with the data from Table 9.2, for which SS_{bet} = .50 and SS_{within} = 6.50, we calculate our degrees of freedom as follows:

$$df_{bet} = 4 - 1$$
$$= 3$$

and

$$df_{within} = 16 - 4$$
$$= 12$$

We are now prepared to obtain the mean squares:

$$MS_{bet} = \frac{.50}{3}$$
$$= .17$$

and

$$MS_{within} = \frac{6.50}{12}$$
$$= .54$$

THE *F* RATIO   As previously noted, the analysis of variance yields an *F* ratio in which variation between groups and variation within groups are compared. We are now ready to specify the degree of each type of variation as measured by mean squares. Therefore, the *F* ratio can be regarded as indicating the size of the between-groups mean square relative to the size of the within-groups mean square, or

$$F = \frac{MS_{bet}}{MS_{within}}$$

For Table 9.2,

$$F = \frac{.17}{.54}$$

$$= .31$$

Having obtained an *F* ratio, we must now determine whether it is large enough to reject the null hypothesis and accept the research hypothesis. Do conservatives, moderates, liberals, and radicals actually differ with respect to child-rearing permissiveness? The larger our calculated *F* ratio (the larger the MS_{bet} and the smaller the MS_{within}), the more likely we will obtain a statistically significant result.

But exactly how do we recognize a significant *F* ratio? Recall that in Chapter 8 our obtained *t* ratio was compared against a table *t* ratio for the .05 level of confidence with the appropriate degrees of freedom. Similarly, we must now interpret our calculated *F* ratio with the aid of Table D in the back of the text. Table D contains a list of significant *F* ratios—*F* ratios that we must obtain in order to reject the null hypothesis at the .05 and .01 levels of confidence. As was the case with the *t* ratio, exactly which *F* value we must get depends on its associated degrees of freedom. Therefore, we enter Table D looking for the two df values, between-groups degrees of freedom and within-groups degrees of freedom. Degrees of freedom associated with the numerator (df_{bet}) have been listed across the top of the page, while degrees of freedom associated with the denominator (df_{within}) have been placed down the left side of the table. The body of Table D presents significant *F* ratios at the .05 and .01 confidence levels.

For the data in Table 9.2, we have found $df_{bet} = 3$ and $df_{within} = 12$. Thus, we move in Table D to the column marked df = 3 and continue down the page from that point until we arrive at the row marked df = 12. By this procedure, we find that a significant *F* ratio at the .05 level of confidence must be at least 3.49, and at the .01 level of confidence must equal or exceed 5.95. Our calculated *F* ratio is only .31. As a result, we have no choice but to *accept* the null hypothesis and attribute our sample mean differences in child-rearing permissiveness to sampling error, rather

Source of Variation	df	SS	MS	F
Between groups	3	.50	.17	.31
Within groups	12	6.50	.54	

than to a true difference in the populations of conservatives, moderates, liberals, and radicals.

The results of our analysis of variance can be presented in a "summary table" such as the one shown in Table 9.3. It has become standard procedure to summarize an analysis of variance in this manner.

An Illustration

To provide a step-by-step illustration of an analysis of variance, suppose we wish to test the hypothesis that IQ varies by social class. Therefore,

Null Hypothesis: The upper, middle, and lower classes do not differ $(M_1 = M_2 = M_3)$ *with respect to IQ.*

Research Hypothesis: The upper, middle, and lower classes differ $(M_1 \neq M_2 \neq M_3)$ *with respect to IQ.*

To investigate this hypothesis, let us say we establish the .05 level of confidence as our criterion of significance. Imagine that we can measure IQ for the members of three samples of social class: upper, middle, and lower. The following IQ scores are assumed to result:

Upper (N = 5)		Middle (N = 5)	
X_1	X^2	X_2	X^2
130	16900	120	14400
125	15625	115	13225
130	16900	115	13225
120	14400	110	12100
122	14884	112	12544
$\Sigma X = 627$	$\Sigma X^2 = 78709$	$\Sigma X = 572$	$\Sigma X^2 = 65494$
$\overline{X}_1 = 125.4$		$\overline{X}_2 = 114.4$	

Lower (N = 5)	
X_3	X^2
110	12100
100	10000
90	8100
100	10000
85	7225
$\Sigma X = 485$	$\Sigma X^2 = 47425$
$\overline{X}_3 = 97.0$	

The step-by-step procedure for testing the statistical significance of the obtained mean difference is as follows.

STEP 1:   Find the Mean for Each Sample

$$\overline{X}_1 = \frac{\Sigma X_1}{N} \qquad \overline{X}_2 = \frac{\Sigma X_2}{N} \qquad \overline{X}_3 = \frac{\Sigma X_3}{N}$$

$$= \frac{627}{5} \qquad\quad = \frac{572}{5} \qquad\quad = \frac{485}{5}$$

$$= 125.4 \qquad\quad = 114.4 \qquad = 97.0$$

Notice that mean differences do exist, the tendency being for IQ scores to increase from lower to middle to upper classes.

STEP 2:   Find the Total Sum of Squares

$$SS_{total} = \Sigma X^2{}_{total} - \frac{(\Sigma X_{total})^2}{N_{total}}$$

$$= (78709 + 65494 + 47425) - \frac{(627 + 572 + 485)^2}{15}$$

$$= 191628 - \frac{(1684)^2}{15}$$

$$= 191628 - \frac{2835856}{15}$$

$$= 191628 - 189057.07$$
$$= 2570.93$$

STEP 3:   Find the Between-Groups Sum of Squares

$$SS_{bet} = \left[\Sigma \frac{(\Sigma X)^2}{N} \right] - \frac{(\Sigma X_{total})^2}{N_{total}}$$

$$= \frac{(627)^2}{5} + \frac{(572)^2}{5} + \frac{(485)^2}{5} - \frac{(1684)^2}{15}$$

$$= \frac{393129}{5} + \frac{327184}{5} + \frac{235225}{5} - \frac{2835856}{15}$$

$$= 78625.8 + 65436.8 + 47045.0 - 189057.07$$
$$= 191107.60 - 189057.07$$
$$= 2050.53$$

STEP 4:   Find the Within-Groups Sum of Squares

$$SS_{within} = SS_{total} - SS_{bet}$$
$$= 2570.93 - 2050.53$$
$$= 520.40$$

or

$$SS_{within} = \Sigma \left[(\Sigma X^2) - \frac{(\Sigma X)^2}{N} \right]$$

$$= \left[78709 - \frac{(627)^2}{5} \right] + \left[65494 - \frac{(572)^2}{5} \right]$$
$$+ \left[47425 - \frac{(485)^2}{5} \right]$$
$$= \left[78709 - \frac{393129}{5} \right] + \left[65494 - \frac{327184}{5} \right]$$
$$+ \left[47425 - \frac{235225}{5} \right]$$
$$= [78709 - 78625.8] + [65494 - 65436.8]$$
$$+ [47425 - 47045.0]$$
$$= 83.2 + 57.2 + 380.0$$
$$= 520.40$$

STEP 5: Find the Between-Groups Degrees of Freedom

$$df_{bet} = k - 1$$
$$= 3 - 1$$
$$= 2$$

STEP 6: Find the Within-Groups Degrees of Freedom

$$df_{within} = N_{total} - k$$
$$= 15 - 3$$
$$= 12$$

STEP 7: Find the Between-Groups Mean Square

$$MS_{bet} = \frac{SS_{bet}}{df_{bet}}$$
$$= \frac{2050.53}{2}$$
$$= 1025.27$$

STEP 8: Find the Within-Groups Mean Square

$$MS_{within} = \frac{SS_{within}}{df_{within}}$$
$$= \frac{520.40}{12}$$
$$= 43.37$$

STEP 9: Obtain the F Ratio

$$F = \frac{MS_{bet}}{MS_{within}}$$
$$= \frac{1025.27}{43.37}$$
$$= 23.64$$

STEP 10:   Compare the Obtained F Ratio with the Appropriate F Ratio in Table D

$$\text{obtained } F \text{ ratio} = 23.64$$
$$\text{table } F \text{ ratio} = 3.88$$
$$\text{df} = \frac{2}{12}$$
$$P = .05$$

As shown in Step 10, to reject the null hypothesis at the .05 level of confidence with $\frac{2}{12}$ degrees of freedom, our calculated F ratio must be at least 3.88. Since we have obtained an F ratio of 23.64, we can reject the null hypothesis and accept the research hypothesis. Specifically, we conclude that the lower, middle, and upper classes actually differ with respect to IQ.

A MULTIPLE COMPARISON OF MEANS

A significant F ratio informs us of an *overall difference* among the groups being studied. If we were investigating a difference between only two sample means, no additional analysis would be needed in order to interpret our result: In such a case, either the obtained difference is statistically significant or it is not, depending on the size of our F ratio. However, when we find a significant F for the differences among three or more means, it may be important to determine exactly where the significant differences lie. For example, in the foregoing illustration, we uncovered statistically significant IQ differences among three social classes. Consider the possibilities raised by this significant F ratio: $\overline{X}_1$ (upper) might differ significantly from $\overline{X}_2$ (middle); $\overline{X}_1$ (upper) might differ significantly from $\overline{X}_3$ (lower); or $\overline{X}_2$ might differ significantly from $\overline{X}_3$ (lower).

As explained earlier in this chapter, obtaining a t ratio for each comparison—$\overline{X}_1$ versus $\overline{X}_2$; $\overline{X}_1$ versus $\overline{X}_3$; $\overline{X}_2$ versus $\overline{X}_3$—would entail a good deal of work and would increase the probability of alpha error as well. Fortunately, a number of other statistical tests have been developed by statisticians for making multiple comparisons after a significant F ratio in order to pinpoint where the significant mean differences lie. We shall introduce Tukey's HSD (honestly significant difference), one of the most useful of the multiple comparison tests.

Tukey's HSD is used only after a significant F ratio has been obtained. By Tukey's method, we compare the difference between any two mean scores against HSD. A mean difference is statistically significant only if it equals or exceeds HSD. By Formula,

$$\text{HSD} = q\alpha \sqrt{\frac{\text{MS}_{\text{within}}}{n}}$$

where

> $q\alpha$ = a table value at a given level of confidence for the maximum number of means being compared
>
> MS_{within} = within-groups mean square (obtained from the analysis of variance)
>
> n = the number of respondents in each group (assumes the same number in each group)

Unlike the t ratio, HSD takes into account the fact that the likelihood of alpha error increases as the number of means being compared increases. Depending on the value of $q\alpha$, the larger the number of means, the more "conservative" HSD becomes in regard to rejecting the null hypothesis. As a result, fewer significant differences will be obtained with HSD than with the t ratio. Moreover, a mean difference is more likely to be significant in a multiple comparison of three means than in a multiple comparison of four or five means.

To illustrate the use of HSD, let us return to a previous example in which social classes were found to differ with respect to IQ. More specifically, we obtained a significant F ratio ($F = 23.64$) for the following differences among upper-, middle-, and lower-class samples:

$$\overline{X}_1 \text{ (upper)} = 125.4$$
$$\overline{X}_2 \text{ (middle)} = 114.4$$
$$\overline{X}_3 \text{ (lower)} = 97.0$$

STEP 1: Construct a Table of Differences Between Ordered Means. For the present data, the rank order of means (from smallest to largest) is 97.0, 114.4, and 125.4. These mean scores are arranged in table form so that the difference between each pair of means is shown in a matrix. Thus, the difference between $\overline{X}_1$ (upper) and $\overline{X}_3$ (lower) is 28.40; the difference between $\overline{X}_1$ (upper) and $\overline{X}_2$ (middle) is 11.0; and the difference between $\overline{X}_2$ (middle) and $\overline{X}_3$ (lower) is 17.4.

	$\overline{X}_3 = 97.0$	$\overline{X}_2 = 114.4$	$\overline{X}_1 = 125.4$
$\overline{X}_3$	—	17.4	28.4
$\overline{X}_2$	—	—	11.0
$\overline{X}_1$	—	—	—

STEP 2: Find $q\alpha$ in Table I. To find $q\alpha$ from Table I in the back of the text, we must have (a) the degrees of freedom (df) for MS_{within}, (b) the maximum number of means (k), and (c) a confidence level,

either .01 or .05. We already know from the analysis of variance that df = 12. Therefore, we enter the left-hand column of Table I until arriving at 12 degrees of freedom. Second, since we are pair-comparing three mean scores, we move across Table I to a maximum number of means (k) equal to three. Assuming a .05 level of confidence, we find that $q.05 = 3.77$.

STEP 3: Find HSD

$$\text{HSD} = q.05 \sqrt{\frac{\text{MS}_{\text{within}}}{n}}$$

$$= 3.77 \sqrt{\frac{43.37}{5}}$$

$$= 3.77 \sqrt{8.67}$$

$$= 3.77(2.94)$$

$$= 11.08$$

STEP 4: Compare HSD Against the Matrix of Mean Differences. To be regarded as statistically significant, any obtained mean difference must equal or exceed HSD. Referring to our matrix of mean differences above, we find that the IQ difference of 28.4 between $\overline{X}_1$ (upper-class) and $\overline{X}_3$ (lower-class) and the IQ difference of 17.4 between $\overline{X}_2$ (middle-class) and $\overline{X}_3$ (lower-class) are greater than HSD = 11.08. As a result, we conclude that these mean differences are statistically significant at the .05 level of confidence. Only the difference of 11.0 between $\overline{X}_1$ and $\overline{X}_2$ does not equal or exceed HSD and is therefore statistically nonsignificant.

REQUIREMENTS FOR THE USE OF THE *F* RATIO

The analysis of variance should be made only after the researcher has considered the following requirements:

1. A comparison between three or more independent means—the *F* ratio is usually employed to make comparisons between three or more means from independent samples. A single sample arranged in a panel design cannot be tested. It is possible, however, to obtain an *F* ratio rather than a *t* ratio when a two-sample comparison is made. For the two-sample case, $F = t^2$, and identical results are obtained.
2. Interval data—to conduct an analysis of variance, we assume that we have achieved the interval level of measurement. Alternatively, categorized or ranked data should not be used.
3. Random sampling—we should have taken our samples at random from a given population of scores.
4. A normal distribution—we assume the sample characteristic we measure to be normally distributed in the underlying population.

SUMMARY

The analysis of variance can be used to make comparisons among three or more sample means. This test yields an F ratio whose numerator represents variation between groups and whose denominator contains an estimate of variation within groups. The sum of squares represents the initial step for measuring variation. However, it is greatly affected by sample size. To overcome this problem, we divide SS_{bet} or SS_{within} by the appropriate degrees of freedom in order to obtain the mean square. The F ratio indicates the size of the between-groups mean square relative to the size of the within-groups mean square. We interpret our calculated F ratio by comparing it against an appropriate F ratio in Table D. On that basis, we decide whether to reject or accept our null hypothesis. After obtaining a significant F, we can determine exactly where the significant differences lie by applying Tukey's method for the multiple comparison of means.

PROBLEMS

1. On the following random samples of social class, test the null hypothesis that neighborliness does not vary by social class. (*Note:* Higher scores indicate greater neighborliness.)

Lower	Working	Middle	Upper
8	7	6	5
4	3	5	2
7	2	5	1
8	8	4	3

2. Test for the significance of differences among means of the following random samples of scores:

Sample 1	Sample 2	Sample 3
2	5	8
1	4	9
3	3	7
3	4	8

3. Test for the significance of differences among means of the following random samples of scores:

Sample 1	Sample 2	Sample 3
12	6	3
6	5	2
8	7	5
7	5	3
6	1	1

4. Test for the significance of differences among means of the following random samples of scores:

Sample 1	Sample 2	Sample 3
5	4	3
5	3	5
4	2	1
3	2	3
6	1	3

5. Conduct a multiple comparison of means by Tukey's method to determine exactly where the significant differences occur in Problem 4 above.
6. Test for the significance of differences among means of the following random samples of scores:

Sample 1	Sample 2	Sample 3	Sample 4
1	3	4	6
1	2	4	6
3	2	2	5
4	1	2	5
2	5	3	4
1	5	3	6

7. Conduct a multiple comparison of means by Tukey's method to determine exactly where the significant differences occur in Problem 6 above.

TERMS TO
REMEMBER

Analysis of variance
Sum of squares
   Within-groups
   Between-groups
   Total
Mean square
F ratio
Multiple comparison of means

10 Chi Square and Other Nonparametric Tests

AS INDICATED IN Chapters 8 and 9, we must ask a good deal of the social researcher who employs a t ratio or an analysis of variance to make comparisons between his samples. Each of these tests of significance has a list of requirements which includes the assumption that the characteristic studied be normally distributed in a specified population. In addition, each test asks for the interval level of measurement, so that a score can be assigned to every sample member. When a test of significance, such as the t ratio or the analysis of variance, requires (1) normality and (2) an interval-level measure, it is referred to as a *parametric test*.[1]

What about the social researcher who cannot employ a parametric test, that is, who either cannot honestly assume normality or whose data are not amenable to an interval-level measure? Suppose, for example, that he or she is working with a skewed distribution, such as annual income, or with data that have been categorized and counted (the nominal level) or ranked (the ordinal level). How does this researcher go about making comparisons between samples without violating the requirements of a particular test?

Fortunately, statisticians have developed a number of *nonparametric* tests of significance—tests whose list of requirements does not include a normal distribution or the interval level of measurement. To understand the important position of nonparametric tests in social research, we must also understand the statistical concept of power. The *power of a test* is the probability of rejecting the null hypothesis when it is actually false and should be rejected.

Power varies from one test to another. The most powerful tests—those that are most likely to reject null when it is false—are tests that have the strongest or most difficult requirements to satisfy. Generally, these are parametric tests such as t or F which assume that interval data have been achieved and that the characteristics being studied are normally

---

[1] This designation is based on the term "parameter," which refers to any characteristic of a population.

distributed in their populations. By contrast, the nonparametric alternatives make less stringent demands and are less powerful tests of significance than their parametric counterparts. As a result, assuming that the null hypothesis is false (and holding constant such other factors as sample size), an investigator is more likely to reject null by the appropriate use of F or t than by a nonparametric alternative.

Understandably, social researchers are anxious to reject the null hypothesis when it is false. As a result, many of them would ideally prefer to employ parametric tests of significance. As previously noted, however, it is often not possible to satisfy the requirements of parametric tests. In the first place, much of the data of social research are at the ordinal or nominal level of measurement. Secondly, we cannot always be sure that the characteristics under study are in fact distributed normally in the population.

When its requirements have been violated, it is not possible to know the power of a statistical test. Therefore, the results of a parametric test whose requirements have gone unsatisfied lack any meaningful interpretation. Under such conditions, many social researchers wisely turn to nonparametric tests of significance.

This chapter introduces some of the best-known nonparametric tests of significance: chi square, the median test, Kruskal-Wallis one-way analysis of variance, and Friedman's two-way analysis of variance.

CHI SQUARE AS A TEST OF SIGNIFICANCE

The most popular nonparametric test of significance in social research is known as *chi square* (χ^2). As we shall see, the χ^2 test is used to make comparisons between two or more samples.

As in the case of the t ratio and analysis of variance, there is a sampling distribution for chi square that can be used to estimate the probability of obtaining a significant chi-square value by chance alone rather than by actual population differences. Unlike the earlier tests of significance, however, chi square is employed to make comparisons between *frequencies* rather than between mean scores. As a result, the null hypothesis for the chi-square test states that the populations do not differ with respect to the frequency of occurrence of a given characteristic, whereas the research hypothesis says that sample differences reflect actual population differences regarding the relative frequency of a given characteristic.

To illustrate the use of chi square for frequency data (or for proportions that can be reduced to frequencies), imagine that we once again have been asked to investigate the relationship between political orientation and child-rearing permissiveness. Rather than *score* liberals and conservatives in terms of their degree of permissiveness, however, we might *categorize* our sample members on a strictly *either-or* basis; that is, we might decide that they are *either* permissive *or* not permissive. Therefore,

Null Hypothesis: The relative frequency of liberals who are permissive is the same as the relative frequency of conservatives who are permissive.

**COMPUTING CHI
SQUARE**

The chi-square test of significance is essentially concerned with the distinction between expected frequencies and obtained frequencies. *Expected frequencies* (f_e) refer to the terms of the null hypothesis, according to which the relative frequency (or proportion) is expected to be the same from one group to another. For example, if 50 percent of the liberals are expected to be permissive, then we also expect 50 percent of the conservatives to be permissive. By contrast, *obtained frequencies* (f_o) refer to the results that we actually obtain when conducting a study and, therefore, may or may not vary from one group to another. *Only if the difference between expected and obtained frequencies is large enough do we reject the null hypothesis and decide that a true population difference exists.*

Continuing with the present example, suppose we were to draw random samples of 20 liberals and 20 conservatives who could be categorized as either permissive or not permissive with respect to child-rearing methods. Table 10.1 shows the obtained frequencies that might result.

The data in Table 10.1 indicate that permissive child-rearing methods were used by 5 out of 20 liberals and 10 out of 20 conservatives. These results can be recast in the form of a 2 × 2 table (2 rows by 2 columns), in which the obtained frequencies are presented for each *cell,* and their expected frequencies are shown in parentheses (see Table 10.2). Note that these expected frequencies are based on the operation of chance alone, assuming therefore that the null hypothesis is correct. Note also that the marginal totals in Table 10.2 (obtained by adding cell frequencies in either direction) are given for the rows (15 and 25) and for the columns (20 and 20). The total number ($N = 40$) can be obtained by adding either the row or column marginals.

Having been given the obtained and expected frequencies for the problem at hand, a chi-square value can now be obtained by the formula

$$\chi^2 = \Sigma \frac{(f_0 - f_e)^2}{f_e}$$

TABLE 10.1
Frequencies obtained in a study of permissiveness by political orientation

Child-Rearing Methods	Political Orientation	
	Liberals f_o	Conservatives f_o
Permissive	5	10
Not permissive	15	10
Total	20	20

TABLE 10.2
*The data from Table 10.1
arranged in a 2 × 2 table*

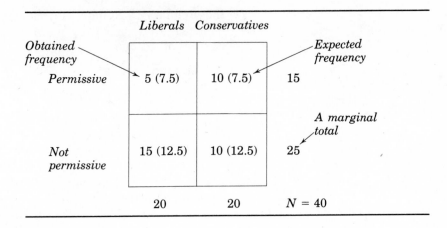

where

$$f_o = \text{the obtained frequency in any cell}$$
$$f_e = \text{the expected frequency in any cell}$$
$$\chi^2 = \text{chi square}$$

According to the formula for χ^2, we must subtract each expected frequency from its corresponding obtained frequency, square the difference, divide by the appropriate expected frequency, and add up these quotients to obtain the chi-square value.

The data in Table 10.2 can be used to illustrate the foregoing procedure:

$$\chi^2 = \frac{(5 - 7.5)^2}{7.5} + \frac{(10 - 7.5)^2}{7.5} + \frac{(15 - 12.5)^2}{12.5}$$
$$+ \frac{(10 - 12.5)^2}{12.5}$$
$$= \frac{(-2.5)^2}{7.5} + \frac{(2.5)^2}{7.5} + \frac{(2.5)^2}{12.5} + \frac{(-2.5)^2}{12.5}$$
$$= \frac{6.25}{7.5} + \frac{6.25}{7.5} + \frac{6.25}{12.5} + \frac{6.25}{12.5}$$
$$= .83 + .83 + .50 + .50$$
$$= 2.66$$

Thus, we learn that $\chi^2 = 2.66$. In order to interpret this chi-square value, we must still determine the appropriate number of degrees of freedom. This can be done for tables having any number of rows and columns by employing the formula

$$df = (r - 1)(c - 1)$$

where

$$r = \text{the number of rows in the table of obtained frequencies}$$

c = the number of columns in the table of obtained frequencies

df = degrees of freedom

Since the obtained frequencies in Table 10.2 form two rows and two columns (2×2),

$$df = (2 - 1)(2 - 1)$$
$$= (1)(1)$$
$$= 1$$

Turning to Table E in the back of the text, we find a list of chi-square values which are significant at the .05 and .01 levels of confidence. For the .05 confidence level, we see that the chi-square value with 1 degree of freedom is 3.84. This is the value that we must equal or exceed before we can reject the null hypothesis. Since our calculated χ^2 is only 2.66 and, therefore, *smaller* than the table value, we must accept the null hypothesis and reject the research hypothesis. The obtained frequencies do not differ enough from the frequencies expected by chance to indicate that actual population differences exist.

FINDING THE EXPECTED FREQUENCIES

The expected frequencies for each cell must reflect the operation of chance under the terms of the null hypothesis. If the expected frequencies are to indicate "sameness" across all samples, they must be proportional to their marginal totals, both for rows and columns.

In order to obtain the expected frequency for any cell, we simply multiply together the column and row marginal totals for a particular cell and divide the product by N. Therefore,

$$f_e = \frac{(\text{row marginal total}) \, (\text{column marginal total})}{N}$$

For the upper-left cell in Table 10.2 (permissive liberals),

$$f_e = \frac{(20)(15)}{40}$$
$$= \frac{300}{40}$$
$$= 7.5$$

Likewise, for the upper-right cell in Table 10.2 (permissive conservatives),

$$f_e = \frac{(20)(15)}{40}$$
$$= \frac{300}{40}$$
$$= 7.5$$

For the lower-left cell in Table 10.2 (not permissive liberals),

$$f_e = \frac{(20)(25)}{40}$$

$$= \frac{500}{40}$$

$$= 12.5$$

For the lower-right cell in Table 10.2 (not permissive conservatives),

$$f_e = \frac{(20)(25)}{40}$$

$$= \frac{500}{40}$$

$$= 12.5$$

As we shall see, the foregoing method for determining f_e can be applied to any chi-square problem for which the expected frequencies must be obtained.

An Illustration

To summarize the step-by-step procedure for obtaining chi square, let us suppose we wanted to study marijuana usage as related to the postgraduation plans of high school students. We might specify our hypotheses as follows:

Null Hypothesis: *The proportion of marijuana smokers among college-oriented high school students is the same as the proportion of marijuana smokers among students who do not plan to attend college.*

Research Hypothesis: *The proportion of marijuana smokers among college-oriented high school students is not the same as the proportion of marijuana smokers among students who do not plan to attend college.*

To test this hypothesis at the .05 level of confidence, say we were to question two random samples from the high school population regarding their use of marijuana: a sample of 21 college-bound students and a sample of 15 students not planning to extend their education beyond high school. Suppose the data in Table 10.3 were

TABLE 10.3
Marijuana smoking among college-oriented and noncollege-oriented students

	College Orientation	
Marijuana Smoking	College f_o	Noncollege f_o
Smokers	15	5
Nonsmokers	6	10
Total	21	15

to result. As shown in the table, 15 out of 21 college-oriented students, but only 5 out of 15 noncollege-oriented students, were marijuana smokers. To find out whether this is a significant difference between college-oriented and noncollege-oriented high school students, we carry out the following step-by-step procedure.

STEP 1:   Rearrange the Data in the Form of a 2 × 2 Table

	College	*Noncollege*	
Smokers	15 ( )	5 ( )	20
Nonsmokers	6 ( )	10 ( )	16
	21	15	$N = 36$

STEP 2:   Obtain the Expected Frequency for Each Cell

15 (11.67)	5 (8.33)	20
6 (9.33)	10 (6.67)	16
21	15	$N = 36$

(upper-left) $f_e = \dfrac{(21)(20)}{36}$

$= \dfrac{420}{36}$

$= 11.67$

(upper-right) $f_e = \dfrac{(15)(20)}{36}$

$= \dfrac{300}{36}$

$= 8.33$

(lower-left) $f_e = \dfrac{(21)(16)}{36}$

$= \dfrac{336}{36}$

$= 9.33$

(lower-right) $f_e = \dfrac{(15)(16)}{36}$

$= \dfrac{240}{36}$

$= 6.67$

STEP 3:   Subtract the Expected Frequencies from the Obtained Frequencies

$$f_o - f_e$$

(upper-left)	15 − 11.67 =	3.33
(upper-right)	5 − 8.33 =	−3.33
(lower-left)	6 − 9.33 =	−3.33
(lower-right)	10 − 6.67 =	3.33

STEP 4: Square This Difference

$$(f_o - f_e)^2$$

(upper-left)	$(3.33)^2 = 11.09$
(upper-right)	$(-3.33)^2 = 11.09$
(lower-left)	$(-3.33)^2 = 11.09$
(lower-right)	$(3.33)^2 = 11.09$

STEP 5: Divide by the Expected Frequency

$$\frac{(f_o - f_e)^2}{f_e}$$

(upper-left) $\dfrac{11.09}{11.67} = .95$

(upper-right) $\dfrac{11.09}{8.33} = 1.33$

(lower-left) $\dfrac{11.09}{9.33} = 1.19$

(lower-right) $\dfrac{11.09}{6.67} = 1.66$

STEP 6: Sum These Quotients to Obtain the Chi-Square Value

$$\Sigma \frac{(f_o - f_e)^2}{f_e}$$

$$.95$$
$$1.33$$
$$1.19$$
$$\underline{1.66}$$
$$\chi^2 = 5.13$$

STEP 7: Find the Degrees of Freedom

$$df = (r - 1)(c - 1)$$
$$= (2 - 1)(2 - 1)$$
$$= (1)(1)$$
$$= 1$$

STEP 8: Compare the Obtained Chi-Square Value with the Appropriate Chi-Square Value in Table E

$$\text{obtained } \chi^2 = 5.13$$
$$\text{table } \chi^2 = 3.84$$
$$df = 1$$
$$P = .05$$

As indicated in Step 8, to reject the null hypothesis at the .05 confidence level with 1 degree of freedom, our calculated chi-square value would have to be 3.84 or larger. Since we have obtained a chi-square value of 5.13, we can reject the null hypothesis and accept the research hypothesis. Our results suggest that the proportion of marijuana smokers is greater among college-bound high school students than among students whose plans do not include attending college.

The step-by-step procedure for chi square just illustrated can be summarized in tabular form as follows:

	f_o	f_e	$f_o - f_e$	$(f_o - f_e)^2$	$\dfrac{(f_o - f_e)^2}{f_e}$
(upper-left)	15	11.67	3.33	11.09	.95
(upper-right)	5	8.33	−3.33	11.09	1.33
(lower-left)	6	9.33	−3.33	11.09	1.19
(lower-right)	10	6.67	3.33	11.09	1.66
					$\chi^2 = 5.13$

A 2 × 2 CHI-SQUARE COMPUTATIONAL FORMULA

We can avoid the lengthy process of calculating the expected frequencies for a 2 × 2 chi-square problem (2 rows by 2 columns) by the use of the following computational formula:

$$\chi^2 = \frac{N(AD - BC)^2}{(A + B)(C + D)(A + C)(B + D)}$$

where

A = obtained frequency in the upper-left cell
B = obtained frequency in the upper-right cell
C = obtained frequency in the lower-left cell
D = obtained frequency in the lower-right cell
N = total number in all cells

We depict cells A, B, C, and D and their marginal totals in a 2 × 2 table as follows:

A	B	$A + B$
C	D	$C + D$
$A + C$	$B + D$	N

To illustrate the use of the computational formula for chi square, let us return to the data in Table 10.3 (marijuana use by college orientation) for which a χ^2 value of 5.13 has already been obtained. We can set up the obtained frequencies for the computational formula as follows:

15	5
A	*B*
C	*D*
6	10

$$\chi^2 = \frac{36[(15)(10) - (5)(6)]^2}{(15 + 5)(6 + 10)(15 + 6)(5 + 10)}$$

$$= \frac{36(150 - 30)^2}{(20)(16)(21)(15)}$$

$$= \frac{36(120)^2}{100800}$$

$$= \frac{36(14400)}{100800}$$

$$= \frac{518400}{100800}$$

$$= 5.14$$

CORRECTING FOR SMALL EXPECTED FREQUENCIES

If the expected frequencies in a 2 × 2 chi-square problem are very small (less than 10 in a cell), the formulas we have learned to this point may yield an inflated chi-square value. Note that this is true only for the *expected* frequencies and not for the frequencies actually obtained in the course of doing research which can be of any size.

In order to reduce the overestimate of chi square and obtain a more conservative result, we apply what is known as *Yates's correction* to the 2 × 2 situation. Using Yates's correction, the difference between obtained and expected frequencies is reduced by .50. Since χ^2 depends on the size of that difference, we also reduce the size of our calculated chi-square value. The corrected chi-square formula for small expected frequencies is located below:

$$\chi^2 = \Sigma \frac{(|f_o - f_e| - .50)^2}{f_e}$$

In the corrected formula above, the straight lines surrounding $f_o - f_e$ indicate that we must reduce the absolute value (ignoring minus signs) of each $f_o - f_e$ by .50.

Let us apply the corrected formula to the data in Table 10.3:

$$\chi^2 = \frac{(|15 - 11.67| - .50)^2}{11.67} + \frac{(|5 - 8.33| - .50)^2}{8.33}$$

$$+ \frac{(|6 - 9.33| - .50)^2}{9.33} + \frac{(|10 - 6.67| - .50)^2}{6.67}$$

$$= \frac{(3.33 - .50)^2}{11.67} + \frac{(-3.33 - .50)^2}{8.33}$$

$$+ \frac{(-3.33 - .50)^2}{9.33} + \frac{(3.33 - .50)^2}{6.67}$$

$$= \frac{(2.83)^2}{11.67} + \frac{(2.83)^2}{8.33} + \frac{(2.83)^2}{9.33} + \frac{(2.83)^2}{6.67}$$

$$= \frac{8.01}{11.67} + \frac{8.01}{8.33} + \frac{8.01}{9.33} + \frac{8.01}{6.67}$$

$$= .69 + .96 + .86 + 1.20$$

$$= 3.71$$

The procedure for applying the corrected chi-square formula can be summarized in tabular form:

f_o	f_e	$|f_o - f_e|$	$|f_o - f_e| - .50$
15	11.67	3.33	2.83
5	8.33	3.33	2.83
6	9.33	3.33	2.83
10	6.67	3.33	2.83

$(|f_o - f_e| - .50)^2$	$\dfrac{(|f_o - f_e| - .50)^2}{f_e}$
8.01	.69
8.01	.96
8.01	.86
8.01	1.20
	$\chi^2 = 3.71$

As shown above, Yates's correction yields a smaller chi-square value ($\chi^2 = 3.71$) than was obtained by means of the uncorrected formula ($\chi^2 = 5.13$). In the present example, our decision regarding the null hypothesis would depend on whether or not we had used Yates's correction. With the corrected formula, we accept the null hypothesis; without it, we reject the null hypothesis.

Yates's correction can also be applied to the computational formula for a 2 × 2 chi square as follows:

$$\chi^2 = \frac{N(|AD - BC| - N/2)^2}{(A + B)(C + D)(A + C)(B + D)}$$

Returning to the data in Table 10.3,

$$\chi^2 = \frac{36[|(15)(10) - (5)(6)| - 36/2]^2}{(15 + 5)(6 + 10)(15 + 6)(5 + 6)}$$

$$= \frac{36(|150 - 30| - 18)^2}{(20)(15)(21)(15)}$$

$$= \frac{36(120 - 18)^2}{100800}$$

$$= \frac{36(102)^2}{100800}$$

$$= \frac{36(10404)}{100800}$$

$$= \frac{374544}{100800}$$

$$= 3.71$$

COMPARING SEVERAL GROUPS

Until now, we have limited our illustrations to the widely employed 2×2 problem. It should be emphasized, however, that chi square is frequently calculated for tables that are larger than 2×2, tables in which several groups or categories are to be compared. The step-by-step procedure for comparing several groups is essentially the same as its 2×2 counterpart. Let us illustrate with a 3×3 problem (3 rows by 3 columns), although any number of rows and columns could be used.

Imagine once more that we were investigating the relationship between political orientation and child-rearing methods. This time, however, say we were able to come up with three random samples: 32 liberals, 30 moderates, and 27 conservatives. Suppose, in addition, we were to categorize the child-rearing methods of our sample members as permissive, moderate, or authoritarian. Therefore,

Null Hypothesis: The relative frequency of permissive, moderate, and authoritarian child-rearing methods is the same for liberals, moderates, and conservatives.

Research Hypothesis: The relative frequency of permissive, moderate, and authoritarian child-rearing methods is not the same for liberals, moderates, and conservatives.

Let us say that we generate the sample differences in child-rearing methods shown in Table 10.4. Therein, we see 7 out of 32 conservatives, 9 out of 30 moderates, and 14 out of 27 liberals could be regarded as permissive in their child-rearing practices.

It must be kept in mind that Yates's correction and the 2×2 computational formula for χ^2 are applicable *only* to the 2×2

TABLE 10.4
Child rearing by political orientation: a 3×3 problem

Child-Rearing Method	Political Orientation		
	Conservative f_o	Moderate f_o	Liberal f_o
Permissive	7	9	14
Moderate	10	10	8
Authoritarian	15	11	5
Total	32	30	27

problem and, therefore, cannot be used for comparing several groups such as in the present 3 × 3 situation. To determine whether or not there is a significant difference in Table 10.4, we must apply the original χ^2 formula which was introduced earlier:

$$\chi^2 = \Sigma \frac{(f_o - f_e)^2}{f_e}$$

The chi-square formula above can be applied to the 3 × 3 problem in the following step-by-step procedure.

STEP 1: Rearrange the Data in the Form of a 3 × 3 Table

Political Orientation

Child-Rearing Methods	*Conservatives*	*Moderates*	*Liberals*	
Permissive	7	9	14	30
Moderate	10	10	8 Obtained frequency	28
Authoritarian	15	11	5	31
	32	30	27	N = 89

Marginal total

STEP 2: Obtain the Expected Frequency for Each Cell

7 (10.79)	9 (10.11)	14 (9.10)	30
10 (10.07)	10 (9.44)	8 (8.49)	28
15 (11.14)	11 (10.45)	5 (9.40)	31
32	30	27	N = 89

(upper-left) $f_e = \frac{(30)(32)}{89} = \frac{960}{89} = 10.79$

(middle-left) $f_e = \frac{(28)(32)}{89} = \frac{896}{89} = 10.07$

$$\text{(lower-left)} \quad f_e = \frac{(31)(32)}{89}$$

$$= \frac{992}{89}$$

$$= 11.14$$

$$\text{(middle-upper)} \quad f_e = \frac{(30)(30)}{89}$$

$$= \frac{900}{89}$$

$$= 10.11$$

$$\text{(middle-middle)} \quad f_e = \frac{(28)(30)}{89}$$

$$= \frac{840}{89}$$

$$= 9.44$$

$$\text{(middle-lower)} \quad f_e = \frac{(31)(30)}{89}$$

$$= \frac{930}{89}$$

$$= 10.45$$

$$\text{(upper-right)} \quad f_e = \frac{(30)(27)}{89}$$

$$= \frac{810}{89}$$

$$= 9.10$$

$$\text{(middle-right)} \quad f_e = \frac{(28)(27)}{89}$$

$$= \frac{756}{89}$$

$$= 8.49$$

$$\text{(lower-right)} \quad f_e = \frac{(31)(27)}{89}$$

$$= \frac{837}{89}$$

$$= 9.40$$

STEP 3: Subtract the Expected Frequencies from the Obtained Frequencies

$$f_o - f_e$$

(upper-left) $7 - 10.79 = -3.79$
(middle-left) $10 - 10.07 = -\ .07$

(lower-left)	15 − 11.14 =	3.86
(upper-middle)	9 − 10.11 =	−1.11
(middle-middle)	10 − 9.44 =	.56
(lower-middle)	11 − 10.45 =	.55
(upper-right)	14 − 9.10 =	4.90
(middle-right)	8 − 8.49 =	− .49
(lower-right)	5 − 9.40 =	−4.40

STEP 4:   Square This Difference

$$(f_o - f_e)^2$$

(upper-left)	$(-3.79)^2 =$	14.36
(middle-left)	$(-.07)^2 =$	.01
(lower-left)	$(3.86)^2 =$	14.90
(upper-middle)	$(-1.11)^2 =$	1.23
(middle-middle)	$(.56)^2 =$	.31
(lower-middle)	$(.55)^2 =$	.30
(upper-right)	$(4.90)^2 =$	24.01
(middle-right)	$(-.49)^2 =$	.24
(lower-right)	$(-4.40)^2 =$	19.36

STEP 5:   Divide by the Expected Frequency

$$\frac{(f_o - f_e)^2}{f_e}$$

(upper-left)	$\dfrac{14.36}{10.79} = 1.33$
(middle-left)	$\dfrac{.01}{10.07} = .00$
(lower-left)	$\dfrac{14.90}{11.14} = 1.34$
(upper-middle)	$\dfrac{1.23}{10.11} = .12$
(middle-middle)	$\dfrac{.31}{9.44} = .03$
(lower-middle)	$\dfrac{.30}{10.45} = .03$
(upper-right)	$\dfrac{24.01}{9.10} = 2.64$
(middle-right)	$\dfrac{.24}{8.49} = .03$
(lower-right)	$\dfrac{19.36}{9.40} = 2.06$

STEP 6:   Sum These Quotients to Obtain the Chi-Square Value

$$\Sigma \frac{(f_o - f_e)^2}{f_e}$$

$$1.33$$
$$.00$$
$$1.34$$
$$.12$$
$$.03$$
$$.03$$
$$2.64$$
$$.03$$
$$\underline{2.06}$$
$$\chi^2 = 7.58$$

STEP 7: Find the Number of Degrees of Freedom

$$\begin{aligned}
\mathrm{df} &= (r - 1)(c - 1) \\
&= (3 - 1)(3 - 1) \\
&= (2)(2) \\
&= 4
\end{aligned}$$

STEP 8: Compare the Obtained Chi-Square Value with the Appropriate Chi-Square Value in Table E

$$\begin{aligned}
\text{obtained } \chi^2 &= 7.58 \\
\text{table } \chi^2 &= 9.49 \\
\mathrm{df} &= 4 \\
P &= .05
\end{aligned}$$

Therefore, we need a chi-square value of at least 9.49 in order to reject the null hypothesis. Since our obtained χ^2 is only 7.58, we must accept the null hypothesis and attribute our sample differences to the operation of chance alone. We have uncovered no statistically significant evidence to indicate that the relative frequency of child-rearing methods differs for liberals, moderates, and conservatives.

REQUIREMENTS FOR THE USE OF CHI SQUARE

Despite the fact that nonparametric tests do not assume a normal distribution in the population, they, too, have a series of requirements that the social researcher must consider if he is to make an intelligent selection among tests of significance. The student will note, however, that requirements for the use of nonparametric tests are generally easier to meet than those for their parametric counterparts such as the *t* ratio or the analysis of variance. Keeping this in mind, let us turn to some of the most important requirements for the use of the chi-square test of significance:

1. A comparison between two or more samples—as illustrated and described in the present chapter, the chi-square test is employed to make comparisons between two or more *independent* samples. This requires that we have at least a

2 × 2 table (at least 2 rows and at least 2 columns). The assumption of independence indicates that chi square cannot be applied to a single sample which has been arranged in a before-after panel design. At least two samples of respondents must be obtained.

2. Nominal data—only frequencies are required.
3. Random sampling—we should have drawn our samples at random from a particular population.
4. The expected cell frequencies should not be too small—exactly how large f_e must be depends on the nature of the problem. For a 2 × 2 problem, no expected frequency should be smaller than 5. In addition, Yates's corrected formula should be used for a 2 × 2 problem in which an expected cell frequency is smaller than 10. For a situation wherein several groups are being compared (say a 3 × 3 or 4 × 5 problem), there is no hard and fast rule regarding minimum cell frequencies, although we should be careful to see that few cells contain less than 5 cases. In any event, the expected frequencies for all cells combined (Σf_e) must always be equal to the obtained frequencies for all cells combined (Σf_o).

THE MEDIAN TEST   Chi square can be applied to any number of independent samples measured at the nominal level. For ordinal data, the *median test* is a simple nonparametric procedure for determining the likelihood that two random samples have been taken from populations with the same median.

To illustrate the procedure for carrying out the median test, suppose an investigator wanted to study male/female reactions to a socially embarrassing situation. To create the embarrassment, the investigator asked 15 men and 12 women with only "average" singing ability individually to sing several songs such as "Love is a Many Splendored Thing" for an audience of "experts." The number of minutes that each subject was willing to continue singing is shown below (a shorter period of time taken to indicate greater embarrassment):

Number of Minutes Sung			
Men	*Women*	*Men*	*Women*
15	12		
18	7	11	9
15	15	10	11
17	16	8	14
17	6	14	9
16	8	9	
10	10	18	
13	6	16	

STEP 1: Find the Median of the Two Samples Combined. By formula,

$$\text{Position of median} = \frac{N+1}{2}$$

$$= \frac{27+1}{2}$$

$$= 14\text{th}$$

The median is the 14th score counting from either end of the distribution arranged in order of size.

To find the median, we arrange all the scores for men and women in consecutive order (without regard for what sample they have come from) and locate the combined median:

18
18
17
17
16
16
16
15
15
15
14
14
13
12 ← Median (the 14th score from either end)
11
11
10
10
10
9
9
9
8
8
7
6
6

STEP 2: Count the Number in Each Sample Falling Above the Median and Not Above the Median (Mdn = 12)

	Men *f*	Women *f*
Above Median	10	3
Not Above Median	5	9
	$N = 27$	

As shown above, the number above and not above the median singing time from each sample, men and women, is represented in a 2×2 frequency table. In the present illustration, 10 of the 15 men, but only 3 of the 12 women, continued singing for a period of time that was greater than the median singing time for the group as a whole.

STEP 3: Perform a Chi-Square Test of Significance. If no sex differences exist with respect to singing time (and therefore social embarrassment), we would expect the same median split within each sample, so that half of the men and half of the women fall above the median. To determine whether the sex differences obtained are statistically significant or merely a product of sampling error, we conduct a χ^2 analysis.

	Men	Women
Above Mdn	10 (*A*)	3 (*B*)
Not Above Mdn	5 (*C*)	9 (*D*)
	$N = 27$	

$$\chi^2 = \frac{N(|AD - BC| - N/2)^2}{(A + B)(C + D)(A + C)(B + D)}$$

$$= \frac{27[|(10)(9) - (3)(5)| - \frac{27}{2}]^2}{(10 + 3)(5 + 9)(10 + 5)(3 + 9)}$$

$$= \frac{27(75 - 13.5)^2}{32760}$$

$$= \frac{102120.75}{32760}$$

$$= 3.12$$

Referring to Table E in the back of the text, we learn that χ^2 must equal or exceed 3.84 (df = 1) in order to be regarded as significant at the .05 level. Since our obtained $\chi^2 = 3.12$, we cannot reject the null hypothesis. There is insufficient evidence to conclude on the basis of our results that men differ from women with respect to their reactions to a socially embarrassing situation.

Requirements for the Use of the Median Test

The following conditions must be satisfied in order to appropriately apply the median test to a research problem.

1. A comparison between two or more independent medians—the median test is employed to make comparisons between two or more medians from independent samples.
2. Ordinal data—to perform a median test, we assume at least the ordinal level of measurement. Nominal data cannot be used.
3. Random sampling—we should have drawn our samples on a random basis from a given population.

FRIEDMAN'S TWO-WAY ANALYSIS OF VARIANCE BY RANKS

In Chapter 8, we introduced a variation of the *t* ratio which could be used to compare the same sample measured twice. In the before-after design, for example, the degree of hostility in a sample of children might be measured both before and after they watch a violent television program.

Friedman's two-way analysis of variance by ranks (χ_r^2) is a nonparametric approach to test differences in a single sample of respondents who have been measured under at least two conditions.

By formula,

$$\chi_r^2 = \frac{12}{Nk(k+1)} \Sigma (\Sigma R_i)^2 - 3N(k+1)$$

where

k = the number of measurements (usually representing the conditions under which respondents are being studied)

N = the total number of cases or respondents

ΣR_i = the sum of ranks for any one measurement (usually representing any one condition being studied)

An Illustration

To illustrate the application of Friedman's two-way analysis of variance, suppose we wish to test the hypothesis that children's hostility varies according to the level of violence in their television programs. In order to study the influence of televised violence, let us imagine that we are able to expose a random sample of ten children to three different levels of televised violence in a program that is essentially the same in all other respects. Let us also say that we have obtained the following hostility scores from these 10 children under each condition of television-viewing (scores range from 20 to 60, higher scores representing greater hostility):

	Condition of Viewing		
Child	Low Violence	Medium Violence	High Violence
A	23	30	32
B	41	45	43
C	36	35	39
D	28	29	35
E	39	41	47
F	25	28	27
G	38	46	51
H	40	47	49
I	45	46	42
J	29	34	38

STEP 1: Rank the Scores of Each Respondent Across All Conditions (in Each Row). In order to conduct Friedman's two-way analysis of variance, we work directly with the ranks for each respondent over all measurements.[2] As shown above, child A's level of

hostility increased from 23 to 30 to 32 as the level of televised violence to which he was exposed increased from low to medium to high. By rank, child A's hostility score was largest (1) under high violence, second-largest (2) under medium violence, and smallest (3) under low violence. Continuing down the list, child B's hostility score was largest (1) under medium violence, second-largest (2) under high violence, and smallest (3) under low violence. Child C was largest (1) under high violence, second-largest (2) under low violence, and smallest (3) under medium violence. The rank-order of each child's three hostility scores has been shown below:

Child	Low Violence	Rank	Medium Violence	Rank	High Violence	Rank
A	23	3	30	2	32	1
B	41	3	45	1	43	2
C	36	2	35	3	39	1
D	28	3	29	2	35	1
E	39	3	41	2	47	1
F	25	3	28	1	27	2
G	38	3	46	2	51	1
H	40	3	47	2	49	1
I	45	2	46	1	42	3
J	29	3	34	2	38	1

STEP 2: Sum the Ranks Under Each Condition (for Each Column). If the null hypothesis is correct—and no significant differences occur between the conditions—we can expect the sum of the ranks across conditions to equal one another (minus sampling error). In the present example, there are three conditions: low, medium, and high televised violence. The ranks for each of these conditions are added as follows:

Child	(Low) Rank	(Medium) Rank	(High) Rank
A	3	2	1
B	3	1	2
C	2	3	1
D	3	2	1
E	3	2	1
F	3	1	2
G	3	2	1
H	3	2	1
I	2	1	3
J	3	2	1
	$\Sigma R = \overline{28}$	$\Sigma R = \overline{18}$	$\Sigma R = \overline{14}$

[2] No tied ranks occurred in the present illustration. In the case of tied ranks (for example, if child A's level of hostility had been precisely the same for two or more levels of violence), follow the procedure for dealing with tied ranks as presented in connection with the rank-order correlation coefficient in Chapter 11.

STEP 3: Substitute in the Formula to Obtain χ_r^2

$$\chi_r^2 = \frac{12}{Nk(k+1)} \Sigma (\Sigma R_i)^2 - 3N(k+1)$$

$$= \frac{12}{(10)(3)(3+1)} (28^2 + 18^2 + 14^2) - 3(10)(3+1)$$

$$= \frac{12}{120} (784 + 324 + 196) - 120$$

$$= .10(1304) - 120$$

$$= 130.4 - 120$$

$$= 10.4$$

STEP 4: Find the Number of Degrees of Freedom

$$\mathrm{df} = k - 1$$
$$= 3 - 1$$
$$= 2$$

STEP 5: Compare χ_r^2 with the Appropriate Chi-Square Value in Table E

$$\text{obtained } \chi_r^2 = 10.4$$
$$\text{table } \chi^2 = 5.99$$
$$\mathrm{df} = 2$$
$$P = .05$$

χ_r^2 is actually a chi-square value derived from the sum of the ranks for all conditions. As a result, we are able to compare our obtained χ_r^2 against the appropriate χ^2 in Table E. With df = 2, we need a chi-square value of at least 5.99 in order to reject the null hypothesis. Since our obtained χ_r^2 is 10.4, we reject the null hypothesis and accept the research hypothesis. We have uncovered evidence to indicate that televised violence does influence hostility in children. There are significant differences in hostility by level of violence.

Requirements for the Use of Friedman's Two-Way Analysis of Variance by Ranks

In order to apply Friedman's two-way analysis of variance, the following conditions must be satisfied:

1. A comparison of a single sample measured under two or more conditions—Friedman's procedure cannot be applied to test differences between independent samples, but assumes that the same sample of respondents has been measured at least twice (or that the members of two or more samples have been matched on relevant variables).

2. Ordinal data—only data capable of being ranked are required.

3. The number of respondents must not be too small—the exact minimal requirement for N depends on the number of conditions (k) to which respondents are to be exposed. For

example, N must equal or exceed 10 when $k = 3$; whereas N must equal or exceed 5 when $k = 4$.

KRUSKAL-WALLIS ONE-WAY ANALYSIS OF VARIANCE BY RANKS

Kruskal-Wallis one-way analysis of variance is a nonparametric alternative to the analysis of variance (F ratio) which can be used to compare several independent samples, but which requires only ordinal-level data. In order to apply the Kruskal-Wallis procedure, we find statistic H as follows:

$$H = \frac{12}{N(N + 1)} \Sigma \left[\frac{(\Sigma R_i)^2}{n} \right] - 3(N + 1)$$

where

N = the total number of cases or respondents
n = the number of cases in a given sample
ΣR_i = sum of the ranks for a given sample

An Illustration

To illustrate the procedure for applying one-way analysis of variance by ranks, consider the possible influence of age on the ability of an individual to find a job. Suppose we study this question by taking random samples from populations of older, middle-aged, and young adults who are given a certain number of days to find a job. Let us say that the following results are obtained:

Number of Days Before Finding a Job		
Older Adults	*Middle-Aged Adults*	*Young Adults*
($n = 7$) 63	($n = 8$) 33	($n = 6$) 25
20	42	31
43	27	6
58	28	14
57	51	18
71	64	13
45	12	
	30	

STEP 1: Rank-Order the Total Group of Scores and Find the Sum of Ranks for Each Sample. All scores must be ranked from lowest to highest (a rank of 1 *must* be assigned to the *smallest* score; 2 to the next smallest score, and so on). In the present illustration, the scores have been ranked from 1 (representing 6 days) to 21 (representing 71 days).[3]

[3] No tied ranks occurred in the present illustration. In the case of tied ranks (for example, if two persons take exactly 24 days to find a job), follow the procedure for dealing with tied ranks as presented in connection with the rank-order correlation coefficient in Chapter 11.

X_1	Rank	X_2	Rank	X_3	Rank
63	19	33	12	25	7
20	6	42	13	31	11
43	14	27	8	6	1
58	18	28	9	14	4
57	17	51	16	18	5
71	21	64	20	13	3
45	15	12	2	$\Sigma R_3 = \overline{31}$	
	$\Sigma R_1 = \overline{110}$	30	10		
			$\Sigma R_2 = \overline{90}$		

STEP 2: Substitute in the Formula to Obtain H

$$H = \frac{12}{N(N+1)} \sum \left[\frac{(\Sigma R_i)^2}{n} \right] - 3(N+1)$$

$$= \left(\frac{12}{21(21+1)} \right) \left(\frac{110^2}{7} + \frac{90^2}{8} + \frac{31^2}{6} \right) - 3(21+1)$$

$$= \left(\frac{12}{462} \right) \left(\frac{12100}{7} + \frac{8100}{8} + \frac{961}{6} \right) - 66$$

$$= (.03)(1728.57 + 1012.50 + 160.17) - 66$$

$$= (.03)(2901.24) - 66$$

$$= 87.04 - 66$$

$$= 21.04$$

STEP 3: Find the Number of Degrees of Freedom

$$\text{df} = k - 1$$
$$= 3 - 1$$
$$= 2$$

STEP 4: Compare H with the Appropriate Chi-Square Value in Table E

$$H = 21.04$$
$$\text{table } \chi^2 = 5.991$$
$$\text{df} = 2$$
$$P = .05$$

To reject the null hypothesis at the .05 level of confidence with 2 degrees of freedom, our calculated H would have to be 5.991 or larger. Since we have obtained an H equal to 21.04, we can reject the null hypothesis and accept the research hypothesis. Our results indicate that there are significant differences, by age, in the amount of time that it takes to find a job.

Requirements for the Use of Kruskal-Wallis One-Way Analysis of Variance

To apply the one-way analysis of variance by ranks, we must consider the following requirements:

1. A comparison of three or more independent samples—one-way analysis of variance cannot be applied to test differences within a single sample of respondents measured more than once.

2. Ordinal data—only data capable of being ranked are required.

3. Each sample must contain at least 6 cases—when there are more than 5 respondents in each group, the significance of H can be determined by means of the appropriate chi-square value in Table E. For testing differences between smaller samples, the reader is referred to special tables in Siegel (1956).

SUMMARY

Statisticians have developed a number of nonparametric tests of significance—tests whose requirements do not include a normal distribution or the interval level of measurement. The best known of the nonparametric tests, chi square, is employed to make comparisons between frequencies rather than between mean scores. When the difference between expected frequencies and obtained frequencies is large enough, we reject the null hypothesis and accept the validity of a true population difference. This is the requirement for a significant chi-square value. Other nonparametric procedures include the median test for determining whether there is a significant difference between the medians of two samples, Friedman's two-way analysis of variance for comparing the same sample measured at least twice, and Kruskal-Wallis one-way analysis of variance by ranks for comparing several independent samples.

PROBLEMS

1. Random samples of males and females were interviewed to determine whether or not they smoked cigarettes. It was found that 15 out of 29 males were smokers; 20 out of 30 females were smokers. Test the null hypothesis that the relative frequency of males who are smokers is the same as the relative frequency of females who are smokers. What do your results indicate?

2. Two groups of students took final exams in statistics. Only one group was given formal course preparation for the exam; the other group read the required text, but never attended lectures. While 22 out of the 30 members of the first group (attendance at lectures) passed the final exam, only 10 out of the 28 members of the second group (no attendance at lectures) passed the exam. Test the null hypothesis that the relative frequency of "attenders" who pass the final exam is the same as the relative frequency of "nonattenders" who pass the exam. What do your results indicate?

3. Applying Yates's correction, conduct a chi-square test of significance for the following 2×2 problem:

16	8
7	11

4. Applying Yates's correction, conduct a chi-square test of significance for the following 2×2 problem:

8	12
10	5

5. Applying Yates's correction, conduct a chi-square test of significance for the following 2×2 problem:

20	14
5	10

6. Conduct a chi-square test of significance for the following 3×3 problem:

20	17	5
15	16	16
4	14	18

7. Conduct a chi-square test of significance for the following 4×2 problem:

25	6
19	10
15	15
8	20

8. Conduct a chi-square test of significance for the following 2×3 problem:

8	10	15
12	10	9

9. Two samples of students were asked to read and then to evaluate a short story written by a new author. One-half of the students were told that the author was a woman, while the other half were told that the author was a man. The following evaluations were obtained (higher scores indicating more favorable evaluations):

X_1 (Told Author Was a Woman)	X_2 (Told Author Was a Man)
6	6
5	8
1	8
1	2
3	5
4	6
3	3
6	8
5	6
5	8
1	2
3	2
5	6
6	8
6	4
3	3

Applying the median test, determine whether there is a significant difference between the medians of these groups. Were student evaluations of the short story influenced by the attributed sex of its author?

10. Applying the median test, determine whether there is a significant difference between the medians of the following samples of scores:

X_1		X_2	
7		4	
8	9	7	3
7	5	3	2
6	9	2	2
7	8	3	6
7	9	4	4
8	7	7	5
9	9	4	4
7	9	5	4
6		6	4
9		2	3

11. The "group harmony and identification" among a sample of 14 children was measured both before and after they had participated in a cooperative classroom task which was designed to make them more dependent on one another for a grade in the course.

The following group identification scores were obtained (higher scores indicating greater group harmony):

Student	Time 1	(Before Cooperative Task)	Time 2	(After Cooperative Task)
A	62		75	
B	51		53	
C	60		62	
D	43		51	
E	49		52	
F	45		46	
G	73		62	
H	66		68	
I	57		55	
J	63		69	
K	43		45	
L	46		45	
M	67		68	
N	61		67	

Applying Friedman's two-way analysis of variance by ranks, determine whether there is a significant difference between Time 1 and Time 2 with respect to group harmony.

12. Applying Friedman's two-way analysis of variance by ranks, determine whether there is a significant difference among Time 1, Time 2, and Time 3 scores of the following sample of 11 respondents:

Respondent	Time 1	Time 2	Time 3
A	60	62	64
B	53	54	50
C	59	65	71
D	65	66	68
E	55	63	61
F	71	74	76
G	57	58	63
H	77	76	79
I	63	65	70
J	54	59	62
K	63	62	65

13. Investigators tested the political alienation among samples of students majoring in liberal arts, engineering, and fine arts. The following results were obtained by sample (higher scores indicating greater alienation):

X_1 (Liberal Arts)	X_2 (Engineering)	X_3 (Fine Arts)
100	101	97
110	90	98
95	92	99
93	100	100

(Continued)

X_1 (Liberal Arts)	X_2 (Engineering)	X_3 (Fine Arts)
106	90	104
102	96	103
	92	

Applying Kruskal-Wallis one-way analysis of variance, determine whether there is a significant difference by college major with respect to level of political alienation.

14. Applying Kruskal-Wallis one-way analysis of variance, determine whether there is a significant difference among the following samples of scores:

X_1	X_2	X_3
125	100	95
100	99	90
122	105	86
127	103	96
115	116	88
129	98	89
130		

TERMS TO REMEMBER

Chi square
Parametric versus nonparametric statistics
Power of a test
Expected frequency
Obtained frequency
Yates's correction
Median test
Friedman's two-way analysis of variance by ranks
Kruskal-Wallis one-way analysis of variance by ranks

11 Correlation

CHARACTERISTICS SUCH AS political orientation, intelligence, and social class *vary* from one respondent to another and, therefore, are referred to as *variables*. In earlier chapters, we have been concerned with establishing the presence or absence of a relationship between any two variables which we shall now label X and Y, for example, between political orientation (X) and child-rearing method (Y), between social class (X) and intelligence (Y), or between college orientation (X) and marijuana use (Y). Aided by the t ratio, analysis of variance, or chi square, we previously sought to discover whether a difference between two or more samples could be regarded as statistically significant—reflective of a true population difference—and not merely the product of sampling error.

STRENGTH OF CORRELATION

Finding that a relationship exists does not indicate much about the degree of association or *correlation* between two variables. Many relationships are statistically significant; few express *perfect* or exact correlation. To illustrate, we know that height and weight are associated, since the taller a person the more he or she tends to weigh. There are numerous exceptions to the rule, however. Some tall people weigh very little; some short people weigh a lot. In the same way, a relationship between college orientation and marijuana smoking does not preclude the possibility of finding many nonsmokers among college-bound students or many smokers among those not planning to attend college.

Correlations actually vary with respect to their *strength*. We can visualize differences in the strength of correlation by means of a *scatter diagram,* a graph that shows the way scores on any two variables X and Y are scattered throughout the range of possible score values. In the conventional arrangement, a scatter diagram is set up so that the X variable is located along the horizontal base line, while the Y variable is measured on the vertical line.

Turning to Figure 11.1, we find two scatter diagrams, each representing the relationship between years of education (X) and income (Y). Figure 11.1(a) depicts this relationship for males, while Figure 11.1(b) represents the relationship for females. Note that each and every point in these scatter diagrams depicts *two* scores, education and income, obtained by *one* respondent. In Figure 11.1(a), for example, we see that a male having 4 years of education earned \$4,000, whereas a male with 13 years of education made \$10,000.

We can say that the strength of the correlation between X and Y increases as the points in a scatter diagram more closely form an imaginary straight line down the center of the graph. Therefore, Figure 11.1(a) (males) represents a stronger correlation than does Figure 11.1(b) (females), although both scatter diagrams indicate that income tends to increase with greater education. Such data would indeed support the view that the income of women (relative to that of men) is less related to the level of education which they attain.

DIRECTION OF CORRELATION

Correlation can often be described with respect to direction as either positive or negative. A *positive correlation* indicates that respondents getting *high* scores on the X variable also tend to get *high* scores on the Y variable. Conversely, respondents who get *low* scores on X also tend to get *low* scores on Y. Positive correlation can be illustrated by the relationship between education and income. As we have previously seen, respondents completing many years of school tend to make large annual incomes, whereas those who complete only a few years of school tend to earn very little annually.

A *negative correlation* exists if respondents who obtain *high* scores on the X variable tend to obtain *low* scores on the Y variable.

FIGURE 11.1
Scatter diagrams representing differences in the strength of the relationship between education and income for males and females

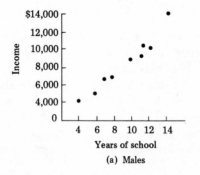

(a) Males

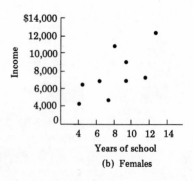

(b) Females

Conversely, respondents achieving *low* scores on X tend to achieve *high* scores on Y. The relationship between education and income would *not* represent a negative correlation, since respondents completing many years of school *do not* tend to make small annual incomes. A more likely example of negative correlation is the relationship between education and prejudice against minority groups. Prejudice tends to diminish as the level of education increases. Therefore, individuals having little formal education tend to hold strong prejudices, whereas individuals completing many years of education tend to be low with respect to prejudice.

CURVILINEAR CORRELATION

A positive or negative correlation represents a type of *straight-line* relationship. Depicted graphically, the points in a scatter diagram tend to form a straight line through the center of the graph. If a positive correlation exists, then the points in the scatter diagram will cluster around the imaginary straight line indicated in Figure 11.2(a). In contrast, if a negative correlation is present, the points in the scatter diagram will surround the imaginary straight line as shown in Figure 11.2(b).

For the most part, social researchers seek to establish straight-line correlation, whether positive or negative. It is important to note, however, that not all relationships between X and Y can be regarded as forming a straight line. There are many *curvilinear* correlations, indicating that one variable increases as the other variable increases until the relationship reverses itself, so that one variable finally decreases while the other continues to increase. That is, a relationship between X and Y that begins as positive becomes negative; a relationship that starts as negative becomes positive. To illustrate a curvilinear correlation, consider the relationship between number of children (family size) and socioeconomic status. As shown in Figure 11.3, the points in the scatter diagram tend to form a U-shaped curve rather than a straight line. Thus, middle-class families have a small number of children: family size (Y) increases as socioeconomic status (X) becomes higher *and* lower.

FIGURE 11.2
Scatter diagrams representing (a) a positive correlation between education and income and (b) a negative correlation between education and prejudice

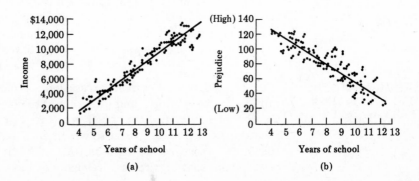

FIGURE 11.3
The relationship between so-
cioeconomic status (X) *and*
family size (Y): *a curvilinear*
correlation

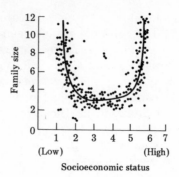

THE CORRELATION COEFFICIENT

The procedure for finding curvilinear correlation lies beyond the scope of this text. Instead, we turn our attention to *correlation coefficients,* which numerically express both the strength and direction of straight-line correlation. Such correlation coefficients generally range between -1.00 and $+1.00$ as follows:

-1.00 ⟵ perfect negative correlation
    ⋮
$-.95$ ⟵ strong negative correlation
    ⋮
$-.50$ ⟵ moderate negative correlation
    ⋮
$-.10$ ⟵ weak negative correlation
    ⋮
$.00$ ⟵ no correlation
    ⋮
$+.10$ ⟵ weak positive correlation
    ⋮
$+.50$ ⟵ moderate positive correlation
    ⋮
$+.95$ ⟵ strong positive correlation
    ⋮
$+1.00$ ⟵ perfect positive correlation

We see, then, that minus numerical values such as -1.00, $-.95$, $-.50$, and $-.10$ signify negative correlation, whereas plus numerical values such as $+1.00$, $+.95$, $+.50$, and $+.10$ indicate positive correlation. Regarding degree of association, the closer to 1.00 in either direction, the greater the strength of the correlation. Since the strength of a correlation is independent of its direction, we can say that $-.10$ and $+.10$ are equal in strength (both are very weak); $-.95$ and $+.95$ have equal strength (both are very strong).

A CORRELATION COEFFICIENT FOR INTERVAL DATA

With the aid of the Pearson correlation coefficient (r), we can determine both the strength and direction of relationship between X and Y variables which have been measured at the interval level. Pearson r reflects the extent to which each sample member obtains the same z score on two variables X and Y. In the case of a positive correlation, both z scores of a respondent have the same sign, either positive or negative, and are located approximately the same distance from the mean of each distribution of scores. Thus, if individual A scores above the mean on X, he also scores above the mean on Y; if individual B scores below the mean on X, he also scores below it on Y. In the case of a negative correlation, the z scores of a respondent have opposite signs, indicating that they are equidistant from, but fall on opposite sides of, their means. If individual A scores above the mean on X, he scores below the mean on Y; if individual B scores below the mean on X, he scores above it on Y. The z-score interpretation of positive and negative correlation has been illustrated in Figure 11.4.

We can now define Pearson r as *the mean of the z-score products for the* X *and* Y *variables.* By formula,

$$r = \frac{\Sigma(z_X z_Y)}{N}$$

where

$r =$ the Pearson correlation coefficient

$z_X =$ an individual's z score on the X variable, equal to $\dfrac{X - \overline{X}}{s_X}$

$z_Y =$ an individual's z score on the Y variable, equal to $\dfrac{Y - \overline{Y}}{s_Y}$

$N =$ the total number of pairs of scores X and Y

FIGURE 11.4
A z-score interpretation of positive versus negative correlation

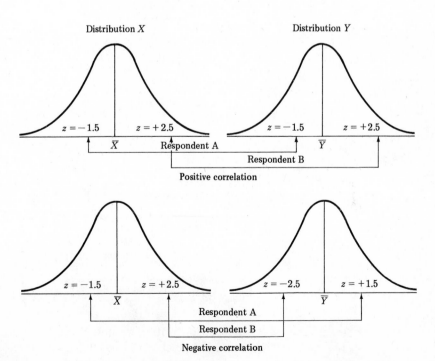

TABLE 11.1
Relationship between respondent's educational level and education of father

		Years of School	
Child	Fathers (X)	Children (Y)	
A	12	12	
B	10	8	
C	6	6	
D	16	11	
E	8	10	
F	9	8	
G	12	11	

To illustrate the application of Pearson r, let us use the above formula to obtain a correlation coefficient for the relationship between the number of years of school completed by the father (X) and the number of years of school completed by his child (Y). The data in Table 11.1 represent this relationship in a random sample of seven respondents.

To apply the Pearson r formula, we must first find $\overline{X}, \overline{Y}, s_X,$ and s_Y as follows:

X	X²	Y	Y²		
12	144	12	144		
10	100	8	64		
6	36	6	36		
16	256	11	121	$\overline{X} = \dfrac{\Sigma X}{N}$	$\overline{Y} = \dfrac{\Sigma Y}{N}$
8	64	10	100		
9	81	8	64		
12	144	11	121	$= \dfrac{73}{7}$	$= \dfrac{66}{7}$
$\Sigma X = 73$	$\Sigma X^2 = 825$	$\Sigma Y = 66$	$\Sigma Y^2 = 650$	$= 10.43$	$= 9.43$

$$s_X = \sqrt{\frac{\Sigma X^2}{N} - \overline{X}^2} \qquad\qquad s_Y = \sqrt{\frac{\Sigma Y^2}{N} - \overline{Y}^2}$$

$$= \sqrt{\frac{825}{7} - (10.43)^2} \qquad = \sqrt{\frac{650}{7} - (9.43)^2}$$

$$= \sqrt{117.86 - 108.78} \qquad = \sqrt{92.86 - 88.92}$$

$$= \sqrt{9.08} \qquad\qquad\qquad = \sqrt{3.94}$$

$$= 3.01 \qquad\qquad\qquad\quad = 1.98$$

For each sample entry, we now find z scores and z score products for the X and Y variables:

	X	$X - \bar{X}$	$\dfrac{X - \bar{X}}{s_X}$	Y	$Y - \bar{Y}$	$\dfrac{Y - \bar{Y}}{s_Y}$	$z_x z_y$
A	12	1.57	.52	12	2.57	1.30	.68
B	10	− .43	−.14	8	−1.43	− .72	.10
C	6	−4.43	−1.47	6	−3.43	−1.73	2.54
D	16	5.57	1.85	11	1.57	.79	1.46
E	8	−2.43	−.81	10	.57	.29	− .24
F	9	−1.43	−.48	8	−1.43	− .72	.34
G	12	1.57	.52	11	1.57	.79	.41
						$\Sigma(z_x z_y) =$	5.29

To illustrate the procedure for obtaining z_X, z_Y, and $z_X z_Y$ above, let us examine the X and Y responses of sample member A. We already know that $X = 10.43$ and $s_X = 3.01$. Since $X - \bar{X} = 12 - 10.43 = 1.57$ for sample member A, we learn that his $z_X = 1.57/3.01 = +.52$. (In other words, A's 12 years of school falls approximately one-half a standard deviation above the mean of the distribution.) Similarly, we know that $\bar{Y} = 9.43$ and $s_Y = 1.98$. Since $Y - \bar{Y} = 12 - 9.43 = 2.57$ for sample member A, we learn that his $z_Y = 2.57/1.98 = +1.30$. (In other words, A's 12 years of school falls approximately one and one-third standard deviations above the mean of this distribution.) To obtain $z_X z_Y$ for A, we multiply his z score $+.52$ by his z score $+1.30$ ($.52 \times 1.30 = .68$). As shown in the right-hand column above, the sum of these z-score products is 5.29.

Plugging into the Pearson formula,

$$r = \frac{\Sigma(z_X z_Y)}{N}$$

$$= \frac{5.29}{7}$$

$$= +.75$$

In the foregoing example, Pearson r equals $+.75$, indicating a fairly strong positive correlation between the educational level attained by children and the educational level of their fathers. That is, respondents whose fathers attained a high level of education also tended to attain a high level of education; respondents whose fathers achieved a low educational level also tended to achieve a low level of education.

A COMPUTATIONAL FORMULA FOR PEARSON r

Computing Pearson r from z scores helps relate the topic of correlation to our earlier discussions of standard scores and the normal curve. However, the z-score formula for Pearson r requires lengthy and time-consuming calculations. Fortunately, there is an alternative formula for Pearson r which works directly with raw scores, thereby eliminating the need to obtain z-score products for X and Y variables. According to the computational formula for Pearson r,

$$r = \frac{N\Sigma XY - (\Sigma X)(\Sigma Y)}{\sqrt{[N\Sigma X^2 - (\Sigma X)^2][N\Sigma Y^2 - (\Sigma Y)^2]}}$$

where

r = the Pearson correlation coefficient
N = the total number of pairs of scores X and Y
X = raw score on the X variable
Y = raw score on the Y variable

In order to illustrate the use of the Pearson r computational formula, let us return to data in Table 11.1 concerning the relationship between the number of years of school completed by the father (X) and the number of years of school completed by his child (Y). To apply the Pearson r formula, we must first obtain X, Y, XY, X^2, and Y^2 as follows:

X	X^2	Y	Y^2	XY
12	144	12	144	144
10	100	8	64	80
6	36	6	36	36
16	256	11	121	176
8	64	10	100	80
9	81	8	64	72
12	144	11	121	132
$\Sigma X = 73$	$\Sigma X^2 = 825$	$\Sigma Y = 66$	$\Sigma Y^2 = 650$	$\Sigma XY = 720$

The Pearson correlation coefficient is equal to

$$r = \frac{7(720) - (73)(66)}{\sqrt{[7(825) - (73)^2][7(650) - (66)^2]}}$$

$$= \frac{5040 - 4818}{\sqrt{(5775 - 5329)(4550 - 4356)}}$$

$$= \frac{222}{\sqrt{(446)(194)}}$$

$$= \frac{222}{\sqrt{86524}}$$

$$= \frac{222}{294.15}$$

$$= +.75$$

Testing the Significance of Pearson r

The Pearson correlation coefficient gives us a precise measure of the strength and direction of correlation in the *sample* being studied. If we have taken a random sample from a specified population, we may still seek to determine whether the obtained association between X and Y exists in the *population* and is not due merely to sampling error.

In order to test the significance of a measure of correlation, we usually set up the null hypothesis that no correlation exists in the population. With respect to the Pearson correlation coefficient, the null hypothesis states that

$$r = 0$$

whereas the research hypothesis says that

$$r \neq 0$$

As was the case in earlier chapters, we test the null hypothesis by selecting a level of confidence such as .05 or .01 and computing an appropriate test of significance. To test the significance of Pearson r, we can compute a t ratio with the degrees of freedom equal to $N - 2$ (N equals the number of pairs of scores). For this purpose, the t ratio can be computed by the formula,

$$t = \frac{r\sqrt{N - 2}}{\sqrt{1 - r^2}}$$

where

$t =$ the t ratio for testing the statistical significance of Pearson r

$N =$ the number of pairs of scores X and Y

$r =$ the obtained Pearson correlation coefficient

Returning to the previous example, we can test the significance of a correlation coefficient equal to $+.754$ between respondent's educational level and the educational level of his father:

$$t = \frac{.754\sqrt{5}}{\sqrt{1 - (.754)^2}}$$

$$= \frac{.754(2.236)}{\sqrt{1 - .569}}$$

$$= \frac{1.69}{\sqrt{.431}}$$

$$= \frac{1.69}{.656}$$

$$= 2.58$$

We turn to Table C in the back of the text and find that a significant t ratio must equal or exceed 2.57 at the .05 level of confidence with 5 degrees of freedom. Since our calculated t ratio ($t = 2.58$) is greater than the required table value, we can reject the null hypothesis that $r = 0$ and accept the research hypothesis that $r \neq 0$. The respondent's educational level and the educational level of the respondent's father are actually associated in the population.

A Simplified Method for Testing the Significance of r

Fortunately, the process of testing the significance of Pearson r as illustrated above has been simplified, so that it becomes unnecessary to actually compute a t ratio. Instead, we turn in the back of the text to Table F, where we find a list of significant values of Pearson r for the .05 and .01 levels of confidence with the number of degrees of freedom ranging from 1 to 90. Directly comparing our calculated value of r with the appropriate table value yields the same result as though we had actually computed a t ratio. If the calculated Pearson correlation coefficient is smaller than the appropriate table value, we must accept the null hypothesis that $r = 0$; if, on the other hand, the calculated r equals or is greater than the table value, we reject the null hypothesis and accept the research hypothesis that a correlation exists in the population.

For illustrative purposes, let us return to our previous example in which a correlation coefficient equal to $+.754$ was tested by means of a t ratio and found to be statistically significant. Turning to Table F in the back of the text, we now find that the value of r must be at least .754 in order to reject the null hypothesis at the .05 level of confidence with 5 degrees of freedom. Hence, this simplified method leads us to the same conclusion as the longer procedure of computing a t ratio.

Correlation: An Illustration

To illustrate the step-by-step procedure for obtaining a Pearson correlation coefficient (r), let us examine the relationship between years of school completed (X) and prejudice (Y) as found in the following sample of ten respondents:

Respondent	Years of School (X)	Prejudice (Y)[a]
A	10	1
B	3	7
C	12	2
D	11	3
E	6	5
F	8	4
G	14	1
H	9	2
I	10	3
J	2	10

[a] Higher scores on the measure of prejudice (from 1 to 10) indicate greater prejudice.

To find Pearson r, we go through the following steps.

STEP 1: Find the Values (1) ΣX, (2) ΣX^2, (3) ΣY, (4) ΣY^2, and (5) ΣXY

Respondent	X	X^2	Y	Y^2	XY
A	10	100	1	1	10
B	3	9	7	49	21
C	12	144	2	4	24
D	11	121	3	9	33
E	6	36	5	25	30
F	8	64	4	16	32
G	14	196	1	1	14
H	9	81	2	4	18
I	10	100	3	9	30
J	2	4	10	100	20
	$\Sigma X = 85$	$\Sigma X^2 = 855$	$\Sigma Y = 38$	$\Sigma Y^2 = 218$	$\Sigma XY = 232$
	(1)	(2)	(3)	(4)	(5)

STEP 2: Plug the Values from Step 1 into the Pearson Correlation Coefficient Formula

$$r = \frac{N\Sigma XY - (\Sigma X)(\Sigma Y)}{\sqrt{[N\Sigma X^2 - (\Sigma X)^2][N\Sigma Y^2 - (\Sigma Y)^2]}}$$

$$= \frac{10(232) - (85)(38)}{\sqrt{[10(855) - (85)^2][10(218) - (38)^2]}}$$

$$= \frac{2320 - 3230}{\sqrt{(8550 - 7225)(2180 - 1444)}}$$

$$= \frac{-910}{\sqrt{(1325)(736)}}$$

$$= \frac{-910}{\sqrt{975200}}$$

$$= \frac{-910}{987.52}$$

$$= -.92$$

Our result indicates a rather strong negative correlation between education and prejudice.

STEP 3: Find the Degrees of Freedom

$$\text{df} = N - 2$$
$$= 10 - 2$$
$$= 8$$

STEP 4: Compare the Obtained Pearson r with the Appropriate Value of Pearson r in Table F

$$\text{obtained } r = -.92$$
$$\text{table } r = .63$$
$$df = 8$$
$$P = .05$$

As indicated above, in order to reject the null hypothesis that $r = 0$ at the .05 level of confidence with 8 degrees of freedom, our calculated value of Pearson r must be at least .63. Since our obtained r equals $-.92$, we reject the null hypothesis and accept the research hypothesis. That is, our result suggests that a correlation between education and prejudice is present in the population from which our sample was taken.

Requirements for the Use of the Pearson Correlation Coefficient

In order to correctly employ the Pearson correlation coefficient as a measure of association between X and Y variables, the following requirements must be taken into account:

1. A straight-line relationship—Pearson r is only useful for detecting a straight-line correlation between X and Y.
2. Interval data—both X and Y variables must be measured at the interval level, so that scores may be assigned to the respondents.
3. Random sampling—sample members must have been drawn at random from a specified population. Otherwise, a test of significance cannot be applied.
4. Normally distributed characteristics—testing the significance of Pearson r requires both X and Y variables to be normally distributed in the population. In small samples, failure to meet the requirement of normally distributed characteristics may seriously impair the validity of Pearson r. However, this requirement is of minor importance when the sample size equals or exceeds 30 cases.

REGRESSION ANALYSIS

Establishing a correlation between two variables can be useful in predicting values of one variable (Y) from knowledge of values of another variable (X). The technique employed in making such predictions is known as *regression analysis*.

Earlier in this chapter, we have seen that the strength of a correlation between X and Y increases as the points in the scatter diagram more closely form an imaginary straight line. We can now identify that line as a *regression line*, a straight line drawn through the scatter diagram, which represents the best possible "fit" for making predictions from X to Y.

Predicting Y from X

Imagine a study concerned with the correlation between the number of years of school completed (X) and annual income (Y), in which we obtain a perfect positive correlation ($r = +1.00$) and the following results for a sample of six respondents:

Respondent	Years of School (X)	Income (Y)
A	18	$30,000
B	6	10,000
C	9	15,000
D	15	25,000
E	12	20,000
F	3	5,000

As shown in Figure 11.5, we can plot the above scores and draw a straight line through them, a regression line which connects the scores of each respondent in the sample. Such a regression line permits the prediction: an individual having 18 years of school will make $30,000; an individual having 3 years of school will make $5,000, and so on.

As pointed out earlier, few of the correlations in social research are perfect, either + 1.00 or − 1.00. This is important since, as a general rule, predictions become more accurate as the size of a correlation becomes larger. For correlations that are less than perfect, we can still construct a prediction or regression line which best "fits" the trend of points in a scatter diagram. This is true, even though all of the points will never lie on that line and our predictions will be less than exact. The regression line for such a less-than-perfect correlation is presented in Figure 11.6.

The Regression Equation

The regression line can be described by means of the formula

$$Y' = r \left(\frac{s_Y}{s_X}\right) X - r \left(\frac{s_Y}{s_X}\right) \overline{X} + \overline{Y}$$

where

Y' = the predicted value of Y (*Note:* It is only a prediction and may vary from Y)

r = the Pearson correlation coefficient for the relationship between X and Y variables

s_Y = sample standard deviation of the distribution of the Y variable

s_X = sample standard deviation of the distribution of the X variable

X = a given value of X

$\overline{X}$ = sample mean of the distribution of the X variable

$\overline{Y}$ = sample mean of the distribution of the Y variable

To illustrate the use of the regression formula to predict values of Y, let us suppose that we have obtained a correlation coefficient equal to + .85 between years of school (X) and annual income (Y).

FIGURE 11.5
A regression line for the relationship between years of school completed (X) *and annual income* (Y) (r = +1.00)

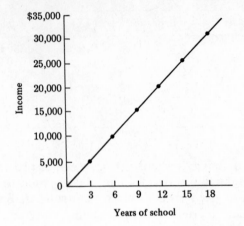

FIGURE 11.5
A regression line for the relationship between years of school completed (X) *and annual income* (Y) (r = +1.00)

Given the data

$$r = +.85$$
$$s_Y = .50$$
$$s_X = .40$$
$$\overline{X} = 10 \text{ years}$$
$$\overline{Y} = \$5000$$

we can now compute the regression equation as follows:

$$Y' = .85 \left(\frac{.5}{.4}\right) X - .85 \left(\frac{.5}{.4}\right) 10 + 5000$$

$$= 1.06X - 1.06(10) + 5000$$
$$= 1.06X - 10.6 + 5000$$
$$= 1.06X + 4989.4$$

To predict the value of Y for each X, we simply "plug in" the values of X. For example: what is the predicted annual income of an individual who has completed 12 years of school? Plugging into the

FIGURE 11.6
A regression line for the relationship between years of school completed (X) *and annual income* (Y) (r < +1.00)

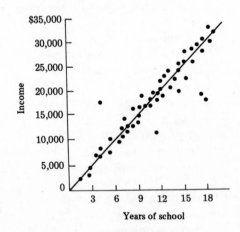

FIGURE 11.6
A regression line for the relationship between years of school completed (X) *and annual income* (Y) (r < +1.00)

regression equation,

$$Y' = 1.06(12) + 4989.4$$
$$= 12.72 + 4989.4$$
$$= 5002.12$$

Therefore, we predict that the annual income of someone having 12 years of school is $5002.12.

In the same way, we can predict that an individual who completes 6 years of school earns $4995.76, or

$$Y' = 1.06(6) + 4989.4$$
$$= 6.36 + 4989.4$$
$$= \$4995.76$$

Regression Analysis: An Illustration

Regression analysis may be further illustrated by returning to examine the relationship between the educational level achieved by fathers (X) and the educational level completed by their children (Y). As noted earlier in this chapter, this relationship in a sample of seven respondents yielded a Pearson correlation coefficient equal to .75:

	Education	
Respondent	*Fathers* (X)	*Respondents* (Y)
A	12	12
B	10	8
C	6	6
D	16	11
E	8	10
F	9	8
G	12	11

We can predict values of Y (child's education) from knowledge of values of X (father's education) by going through the following steps:

STEP 1: Find the Pearson Correlation Coefficient

$$r = \frac{N\Sigma XY - (\Sigma X)(\Sigma Y)}{\sqrt{[N\Sigma X^2 - (\Sigma X)^2][N\Sigma Y^2 - (\Sigma Y)^2]}}$$

$$= \frac{7(720) - (73)(66)}{\sqrt{[7(825) - (73)^2][7(650) - (66)^2]}}$$

$$= \frac{5040 - 4818}{\sqrt{(5775 - 5329)(4550 - 4356)}}$$

$$= \frac{222}{\sqrt{86524}}$$

$$= \frac{222}{294.15}$$

$$= +.754$$

STEP 2:   Obtain the Sample Mean for X and Y

$$\overline{X} = \frac{\Sigma X}{N} \qquad\qquad \overline{Y} = \frac{\Sigma Y}{N}$$

$$= \frac{73}{7} \qquad\qquad = \frac{66}{7}$$

$$= 10.43 \qquad\qquad = 9.43$$

STEP 3:   Obtain the Sample Standard Deviation for X and Y

$$s_X = \sqrt{\frac{\Sigma X^2}{N} - \overline{X}^2} \qquad\qquad s_Y = \sqrt{\frac{\Sigma Y^2}{N} - \overline{Y}^2}$$

$$= \sqrt{\frac{825}{7} - (10.43)^2} \qquad\qquad = \sqrt{\frac{650}{7} - (9.43)^2}$$

$$= \sqrt{117.86 - 108.79} \qquad\qquad = \sqrt{92.86 - 88.93}$$

$$= \sqrt{9.07} \qquad\qquad\qquad = \sqrt{3.93}$$

$$= 3.01 \qquad\qquad\qquad\quad = 1.98$$

STEP 4:   Plug the Values from Steps 1, 2, and 3 into the Regression Equation

$$Y' = r\left(\frac{s_Y}{s_X}\right) X - r\left(\frac{s_Y}{s_X}\right) \overline{X} + \overline{Y}$$

$$= .75\left(\frac{1.98}{3.01}\right) X - .75\left(\frac{1.98}{3.01}\right) 10.43 + 9.43$$

$$= .75(.66)X - .75(.66)10.43 + 9.43$$

$$= .50X - 5.22 + 9.43$$

$$= .50X + 4.21$$

STEP 5:   Determine the Value of Y' for Values of X

[*Examples*]

1. For a respondent whose father completed 16 years of education:

$$Y' = .50X + 4.21$$

$$= .50(16) + 4.21$$

$$= 8.0 + 4.21$$

$$= 12.21$$

2. For a respondent whose father completed 6 years of education:

$$Y' = .50X + 4.21$$

$$= .50(6) + 4.21$$

$$= 3.0 + 4.21$$

$$= 7.21$$

Conclusion: We can predict that respondents whose fathers have completed 16 years of school will have completed 12.21 years of education; respondents whose fathers have completed 6 years of school will have completed 7.21 years of education.

CORRELATION COEFFICIENT FOR ORDINAL DATA

To this point, we have presented Pearson r, a correlation coefficient for application to data that can be scored at the interval level of measurement. We turn now to the problem of finding the degree of association for ordinal data—data that have been ranked or ordered with respect to the presence of a given characteristic.

To take an example from social research, consider the relationship between socioeconomic status and amount of time spent watching television. Imagine that a sample of eight respondents could be ranked as follows:

Respondent	Socioeconomic Status (X) Rank		Time Spent Watching TV (Y) Rank	
Max	1	highest	2	most
Flory	2	in socio-	1	time
Jack	3	economic	3	watching
Min	4	status	5	TV
Lenny	5		4	
Linda	6		8	
Carol	7		6	
Margie	8		7	

As shown above, Max ranked first with respect to socioeconomic status, but second in regard to the amount of time spent watching television; Flory's position was second with respect to socioeconomic status, and first in terms of time spent watching TV, and so on.

In order to determine the degree of association between socioeconomic status and amount of time watching television, we apply Spearman's *rank-order correlation coefficient* (r_s). By formula,

$$r_s = 1 - \frac{6\Sigma D^2}{N(N^2 - 1)}$$

where

r_s = the rank-order correlation coefficient
D = rank difference between X and Y variables
N = the total number of cases

We set up the present example as shown in Table 11.2.

TABLE 11.2
The relationship between socioeconomic status and time spent watching television

Respondent	Socioeconomic Status X	Time Watching Television Y	D	D^2
1	1	2	−1	1
2	2	1	1	1
3	3	3	0	0
4	4	5	−1	1
5	5	4	1	1
6	6	8	−2	4
7	7	6	1	1
8	8	7	1	1
				$\Sigma D^2 = \overline{10}$

Applying the rank-order correlation coefficient to the data in Table 11.2,

$$r_s = 1 - \frac{6(10)}{8(64-1)}$$

$$= 1 - \frac{60}{8(63)}$$

$$= 1 - \frac{60}{504}$$

$$= 1 - .12$$

$$= +.88$$

Therefore, we find a strong positive correlation ($r_s = +.88$) between socioeconomic status and time spent watching television: respondents having high socioeconomic status tend to watch a good deal of television; respondents who have low socioeconomic status tend to spend little time watching television.

Dealing with Tied Ranks

In actual practice, it is not always possible to rank or order our respondents avoiding ties at each and every position. We might find for instance, that two or more respondents spend exactly the same amount of time in front of the television set, that the academic achievement of two or more students is indistinguishable, or that several respondents have the same IQ score.

To illustrate the procedure for obtaining a rank-order correlation coefficient in the case of tied ranks, let us say we are interested in determining the degree of association between position in a graduating class and IQ. Suppose also we are able to rank a sample of ten graduating seniors with respect to their class position and to obtain their IQ scores as follows:

Respondent	Class Standing X	IQ Y
Jim	10 ←——(last)	110
Tracy	9	90
Steffie	8	104
Mike	7	100
David	6	110
Kenny	5	110
Mitchell	4	132
Maxie	3	115
Cori	2	140
	1 ←——(first)	140

Before following the standard procedure for obtaining a rank order correlation coefficient, let us first rank the IQ scores of our ten graduating seniors:

Respondent	IQ	IQ Rank	
Jim	110	7	
Tracy	90	10	
Steffie	104	8	positions 5, 6,
Mike	100	9	and 7 are tied
David	110	6	
Kenny	110	5	
Mitchell	132	3	
Maxie	115	4	
Cori	140	2	positions 1 and
Sue	140	1	2 are tied

As shown above, Cori and Sue received the highest IQ scores and are, therefore, tied for the first and second positions. Likewise, Kenny, David, and Jim achieved an IQ score of 110, which places them in a three-way tie for the fifth, sixth, and seventh positions.

To determine the exact position in the case of ties, we must *add the tied ranks and divide by the number of ties*. Therefore, the position of a 140 IQ, which has been ranked as 1 and 2, would be the "average" rank

$$\frac{1 + 2}{2} = 1.5$$

In the same way, we find that the position of an IQ score of 110 is

$$\frac{5 + 6 + 7}{3} = 6.0$$

Having found the ranked position of each IQ score, we can proceed to set up the problem at hand as shown in Table 11.3.

Respondent	Class Standing (X)	IQ (Y)	X − Y = D	D²
1	10	6	4.0	16.00
2	9	10	−1.0	1.00
3	8	8	0	0
4	7	9	−2.0	4.00
5	6	6	0	0
6	5	6	−1.0	1.00
7	4	3	1.0	1.00
8	3	4	−1.0	1.00
9	2	1.5	.5	.25
10	1	1.5	− .5	.25
				$\Sigma D^2 = \overline{24.50}$

We obtain the rank-order correlation coefficient for the problem in Table 11.3 as follows:

$$r_s = 1 - \frac{6(24.50)}{10(100-1)}$$

$$= 1 - \frac{147}{990}$$

$$= 1 - .15$$

$$= +.85$$

The resultant rank-order coefficient indicates a rather strong *positive* correlation between class standing and IQ. That is, students having *high* IQ scores tend to rank *high* in their class; students who have *low* IQ scores tend to rank *low* in their class.

Testing the Significance of the Rank-Order Correlation Coefficient

How do we go about testing the significance of a rank-order coefficient? For example: How can we determine whether the obtained correlation of + .85 between class standing and IQ can be generalized to a larger population? To test the significance of a computed r_s we simply turn in the back of the text to Table G, where we find the significant values of the rank-order coefficient of correlation for the .05 and .01 confidence levels. Notice that we refer directly to the number of pairs of scores (N) rather than to a particular number of degrees of freedom. In the present case, $N = 10$ and a significant r_s must equal or exceed .648. We therefore reject the null hypothesis that $r_s = 0$ and accept the research hypothesis that class standing and IQ are actually related in the population from which our sample was drawn.

Rank-Order Correlation: An Illustration

We can summarize the step-by-step procedure for obtaining the rank-order correlation coefficient with reference to the relationship between the degree of participation in voluntary associations and number of close friends. This relationship is indicated in the following sample of five respondents:

Respondent	*Voluntary Association Participation (X)* Rank	*Number of Friends (Y)*
A	1 ◄── participates	6
B	2 most	4
C	3	6
D	4	2
E	5 ◄── participates least	2

To determine the degree of association between voluntary association participation and number of friends, we carry through the following steps.

STEP 1: Rank Respondents on the *X* and *Y* Variables. As shown above, we rank respondents with respect to *X*, participation in voluntary associations, assigning the rank of 1 to the respondent who participates most and the rank of 5 to the respondent who participates least.

We must also rank the respondents in terms of *Y*, the number of their friends. In the present example, we have instances of tied ranks as shown below:

Number of Friends (Y)	Rank
6	1
4	3 tied for first
6	2 and second
2	4 tied for fourth
2	5 and fifth

To convert tied ranks, we take an "average" of the tied positions:

For first and second positions: $\frac{1+2}{2} = 1.5$

For fourth and fifth positions: $\frac{4+5}{2} = 4.5$

Therefore,

X	Y
1	1.5
2	3.0
3	1.5
4	4.5
5	4.5

STEP 2:   Find ΣD^2. We must find the difference between X and Y ranks (D), square each difference (D^2), and sum these squares (ΣD^2):

X	Y	D	D²
1	1.5	− .5	.25
2	3.0	−	1.00
3	1.5	.5	2.25
4		.5	.25
	4.0	.5	.25
			$\Sigma D^2 = 4.00$

STEP 3:   Plug the Result of Step 2 into the Formula for the Rank-Order Correlation Coefficient

$$r_s = 1 - \frac{6\Sigma D^2}{N(N^2 - 1)}$$

$$= 1 - \frac{6(4)}{5(24)}$$

$$= 1 - \frac{24}{120}$$

$$= 1 - .20$$

$$= + .80$$

STEP 4:   Compare the Obtained Rank-Order Correlation Coefficient with the Appropriate Value of r_s in Table G

$$\text{obtained } r_s = .80$$
$$\text{table } r_s = 1.00$$
$$N = 5$$
$$P = .05$$

Turning to Table G in the back of the text, we learn that a correlation coefficient of 1.00 (perfect correlation) is necessary to reject the null hypothesis at the .05 level of confidence with a sample size of 5. Therefore, although we have uncovered a strong positive correlation between voluntary association participation and number of friends, we must still accept the null hypothesis that $r_s = 0$. Our result cannot be generalized to the population from which our sample was taken.

Requirements for the Use of the Rank-Order Correlation Coefficient

The rank-order correlation coefficient should be employed when the following conditions can be satisfied:

1. A straight-line correlation—the rank-order coefficient detects straight-line relationships between X and Y.
2. Ordinal data—both X and Y variables must be ranked or ordered.
3. Random sampling—sample members must have been taken at random from a larger population.

GOODMAN'S AND KRUSKAL'S GAMMA

Correlation can be viewed in terms of the degree to which the values of one variable can be predicted or guessed from knowledge of the values of the other variable. This can be seen quite directly in Goodman's and Kruskal's *gamma* (*G*), an alternative to the rank-order correlation coefficient that many social researchers prefer for measuring the degree of association between ordinal-level variables.

The basic formula for *gamma* is

$$G = \frac{\Sigma f_a - \Sigma f_i}{\Sigma f_a + \Sigma f_i}$$

where

f_a = the frequency of agreements
f_i = the frequency of inversions

Agreements and inversions can be understood as expressing the direction of correlation between X and Y variables. Perfect agreement indicates a perfect positive correlation ($+1.00$): all individuals being studied have been ranked in exactly the same order on both variables. As shown below, an individual who ranks first on X also ranks first on Y; an individual who ranks second on X also ranks second on Y; and so on.

	Rank On	
Individuals	*X*	*Y*
A	1	1
B	2	2
C	3	3
D	4	4
E	5	5
F	6	6

By contrast, perfect inversion indicates a perfect negative correlation (-1.00), so that the individuals being studied are ranked in exactly reverse order on two variables. Thus, an individual who ranks first on X ranks last on Y; an individual who ranks second on X ranks second-to-last on Y; and so on.

	Rank On	
Individuals	*X*	*Y*
A	1	6
B	2	5
C	3	4
D	4	3
E	5	2
F	6	1

Whenever perfect agreement or perfect inversion occurs, it becomes possible to predict with total accuracy an individual's rank on one variable from knowledge of his rank on the other variable. In the case of perfect agreement, for example, we know that a person who ranks third on X also ranks third on Y. Since perfect correlation rarely occurs in the practice of social research, however, our ability to make accurate predictions about one variable based on knowledge of another must depend on the *amount* of agreement or inversion in the rank-ordering of individuals on the two variables.

The Gamma Coefficient: An Illustration

To illustrate the use of *gamma,* let us say we were studying the size of the black population in metropolitan areas of the United States as related to their level of job discrimination. Such a study might be carried out, for example, by analyzing the population and income data presently available from the U.S. Bureau of the Census.

Suppose that we were able to rank-order the six largest metropolitan areas of the United States with respect to both the size of their black population (X) and their level of discrimination (Y), as follows:

Metropolitan Area	Size of Black Population (X)	Level of Job Discrimination (Y)
A	6	4
B	1	2
C	2	3
D	5	5
E	4	6
F	3	1

Thus, we see that Metropolitan Area A had the smallest number of blacks and was fourth highest with respect to discrimination; Metropolitan Area B had the largest black population and was second with respect to discrimination; and so on.

STEP 1: Rearrange the Data so that the X Variable Is Perfectly Ordered from Highest to Lowest. To determine the degree of association between size of black population and job discrimination, we first set up the data in a table in which the X variable (in this case, size of black population) has been perfectly ordered from first (1) to last (6) and the Y variable (in this case, level of discrimination) is left unordered. The frequency of agreements and inversions in the unordered column (the Y variable) indicates how much this column of ranks differs from a perfectly ordered ranking, either positive (1, 2, 3, 4, 5, 6) or negative (6, 5, 4, 3, 2, 1):

Metropolitan Area	Size of Black Population (X)	Level of Job Discrimination (Y)
B	1	2
C	2	3
F	3	1
E	4	6
D	5	5
A	6	4

STEP 2: Obtain Frequency of Agreements. To find the frequency of agreements (f_a), we begin with the top-most rank in the Y column (Metropolitan Area B in the above table). For each rank, we count *the number of ranks above it in the table that are smaller in numerical value.* The number of ranks that occur above the top-most rank is always zero (since there are no ranks above the top-most entry in the table). As a result, we enter a zero in the agreements column for Metropolitan Area B. Moving to the second-listed rank in the Y column (Metropolitan Area C), we count the number of ranks falling above it in the table that are smaller in numerical value. We see that only the rank of 2 falls above that for Metropolitan Area C. And, since this rank is smaller than 3, we enter a 1 in the agreements column. Continuing to the next-listed rank (Metropolitan Area F), we find a rank of 1. Since the ranks above it (3 and 2) are larger than 1, we enter a zero in the agreements column. Again moving downward in the Y column to Metropolitan Area E, we count the number of ranks above it that are smaller than 6. Since all three ranks above (1, 3, 2) are smaller, we enter a 3 in the agreements column. We go on to the remaining ranks in column Y and repeat the procedure of counting and entering agreements.

Metropolitan Area	Size of Black Population (X)	Level of Job Discrimination (Y)	Agreements
B	1	2	0
C	2	3	1
F	3	1	0
E	4	6	3
D	5	5	3
A	6	4	3

STEP 3: Obtain Frequency of Inversions. To find the frequency of inversions, we again begin with the top-most entry in the Y column (Metropolitan Area B). This time, however, for each rank, we count *the number of ranks falling above it that are larger in numerical value.* Beginning with the top-most rank, we again see that there are no ranks occurring above and enter a zero in the inversions column. Continuing to the second-listed rank in the Y column

(Metropolitan Area C), we count the number of ranks falling above 3 that are larger in value. Only the rank of 2 falls above that for Metropolitan Area C. Since this rank is smaller, not larger, than 3, we enter a zero in the inversions column. Moving down to the next-listed rank (Metropolitan Area F), we find a rank of 1. Since both of the ranks above it (3 and 2) are larger than 1, we enter a 2 in the inversions column. Moving downward once more, we encounter a rank of 6 for Metropolitan Area E. Since none of the ranks above it (1, 3, 2) is larger than 6, we enter a zero in the inversions column. We then continue to the remaining ranks and repeat the procedure of counting and entering inversions.

Metropolitan Area	Size of Black Population (X)	Level of Job Discrimination (Y)	Inversions
B	1	2	0
C	2	3	0
F	3	1	2
E	4	6	0
D	5	5	1
A	6	4	2

STEP 4: Obtain Σf_a and Σf_i. Once all agreements and inversions have been counted, we add up the agreements (Σf_a) and the inversions (Σf_i) as shown below:

	Agreements	Inversions
B	0	0
C	1	0
F	0	2
E	3	0
D	3	1
A	3	2
	$\Sigma f_a = \overline{10}$	$\Sigma f_i = \overline{5}$

STEP 5: "Plug" Σf_a and Σf_i into the Gamma Formula

$$G = \frac{\Sigma f_a - \Sigma f_i}{\Sigma f_a + \Sigma f_i}$$

$$= \frac{10 - 5}{10 + 5}$$

$$= \frac{5}{20}$$

$$= +.25$$

A *gamma* coefficient equal to +.25 indicates the presence of a weak positive correlation. This is a correlation based on a dominance of agreements: There is 25 percent greater agreement than inversion between the size of the black population and job discrimination.

Dealing with Tied Ranks

As we saw in connection with the rank-order correlation coefficient, it is not always possible to avoid tied ranks at the ordinal level of measurement. In fact, social researchers frequently work with crude ordinal measures which produce large numbers of tied ranks. When a very large number of ties occur, gamma's simple computational procedures make it an especially useful measure of association. The basic formula for gamma is employed for tied ranks, but the frequencies of agreements and inversions are computed in a somewhat different way.

Let us illustrate the procedure for obtaining a gamma coefficient with tied ranks. Suppose that a researcher wanting to examine the relationship between social class and voluntary association membership obtained the following data from a questionnaire study of 80 city residents: among 29 upper-class respondents, 15 were "high," 10 were "medium," and 4 were "low" with respect to voluntary association membership; among 25 respondents who were middle-class, 8 were "high," 10 were "medium," and 7 were "low" with respect to voluntary association membership; and among 26 lower-class respondents, 7 were "high," 8 were "medium," and 11 were "low" with respect to voluntary association membership. Notice that tied ranks occur at every position. For instance, there were 29 respondents who tied at the rank of upper social class, the highest rank on the X variable.

STEP 1: Rearrange the Data in the Form of a Frequency Table

Voluntary Association Membership (Y)	Social Class (X)		
	Upper	*Middle*	*Lower*
High	15	8	7
Medium	10	10	8
Low	4	7	11
	29	25	26
		N = 80	

Notice that the table above is a 3 × 3 frequency table containing 9 cells (3 rows × 3 columns = 9). To assure that the sign of the gamma coefficient is accurately depicted as either positive or negative, the X variable in the columns must always be arranged in decreasing order from left to right. In the table, for example, social class decreases—upper, middle, lower—from left to right columns. Similarly, the Y variable in the rows must decrease from top to bottom. In the table above, voluntary association membership decreases—high, medium, low—from top to bottom rows.

Decision Making

STEP 2: Obtain Σf_a. To find Σf_a, begin with the cell ($f = 15$) in the upper left-hand corner. Multiply this number by the sum of all numbers that fall *below and to the right of it*. Reading from left to right, we see that all frequencies below *and* to the right of 15 are 10, 8, 7, and 11. Now repeat this procedure for all cell frequencies that have cells below and to the right of them. Working from left to right in the table:

Upper-class/high membership	$15(10 + 8 + 7 + 11) = 15(36) = 540$
Middle-class/high membership	$8(8 + 11) = 8(19) = 152$
Upper-class/medium membership	$10(7 + 11) = 10(18) = 180$
Middle-class/medium membership	$10(11) = 110$

(Note that none of the other cell frequencies in the table—7 in the top row, 8 in the second row, and 4, 7, and 11 in the bottom row—has cells below *and* to the right.)

Σf_a is the sum of the products obtained above. Therefore,

$$\Sigma f_a = 540 + 152 + 180 + 110$$
$$= 992$$

STEP 3: Obtain Σf_i. To obtain Σf_i, reverse the procedure for finding agreements and begin in the upper *right*-hand corner of the table. This time, each number is multiplied by the sum of all numbers that fall *below and to the left of it*. Reading from right to left, we see that frequencies below *and* to the left of 7 are 10, 10, 7, and 4. As before, repeat this procedure for all frequencies having cells below and to the right of them.

Working from right to left,

Lower-class/high membership	$7(10 + 10 + 7 + 4) = 7(31) = 217$
Middle-class/high membership	$8(10 + 4) = 8(14) = 112$
Lower-class/medium membership	$8(7 + 4) = 8(11) = 88$
Middle-class/medium membership	$10(4) = 40$

(Note that none of the other cell frequencies in the table—15 in the top row, 10 in the middle row, 11, 7, and 4 in the bottom row—has cells below and to the left.)

Σf_i is the sum of the products computed above. Therefore,

$$\Sigma f_i = 217 + 112 + 88 + 40$$
$$= 457$$

STEP 4: "Plug" the Results of Steps 2 and 3 into the Formula for Gamma

$$G = \frac{\Sigma f_a - \Sigma f_i}{\Sigma f_a + \Sigma f_i}$$

$$= \frac{992 - 457}{992 + 457}$$

$$= \frac{535}{1449}$$

$$= +.37$$

A gamma coefficient of $+.37$ indicates a moderately weak positive correlation between social class and voluntary association membership. Our result suggests a correlation based on a dominance of agreements: There is 37 percent greater agreement than inversion between social class and voluntary association membership. (Note that a gamma coefficient of $-.37$ would have indicated instead a moderately weak *negative* correlation based on a dominance of *inversions*.)

Testing the Significance of Gamma

In order to test the null hypothesis that X and Y are not associated in the population, we convert our calculated G to a z score by the following formula:

$$z = G \sqrt{\frac{\Sigma f_a - \Sigma f_i}{N(1 - G^2)}}$$

where

G = the calculated gamma coefficient
f_a = frequency of agreements
f_i = frequency of inversions

In the foregoing illustration, we found that $G = .37$ for the correlation between social class and voluntary association membership. To test the significance of our finding, we substitute in the formula as follows:

$$z = (.37) \sqrt{\frac{992 - 457}{80(1 - .37^2)}}$$

$$= (.37) \sqrt{\frac{535}{80(.86)}}$$

$$= (.37) \sqrt{\frac{535}{68.80}}$$

$$= (.37)\sqrt{7.78}$$

$$= (.37)(2.79)$$

$$= 1.03$$

Turning to Table B in the back of the text, we see that z must equal or exceed 1.96 to reject null at the .05 level of confidence.

Since our calculated z ($z = 1.03$) is smaller than the required table value, we must accept the null hypothesis that $G = 0$ and reject the research hypothesis that $G = 0$. Our obtained correlation cannot be generalized to the population from which our sample was drawn.

Requirements for the Use of Gamma

The following requirements must be taken into account in order to employ gamma as a measure of association:

1. A straight-line correlation—gamma detects straight-line relationships between X and Y.
2. Ordinal data—both X and Y variables must be ranked or ordered.
3. Random sampling—to test the null hypothesis ($G = 0$), sample members must have been taken on a random basis from some specified population.

CORRELATION COEFFICIENT FOR NOMINAL DATA ARRANGED IN A 2 × 2 TABLE

In the last chapter, we were introduced to a test of significance for frequency data known as chi square. By a simple extension of the chi-square test, we can now determine the degree of association between variables at the nominal level of measurement.

Let us take another look at the null hypothesis that

the proportion of marijuana smokers among college-oriented high school students is the same as the proportion of marijuana smokers among students who do not plan to attend college.

In Chapter 10, this null hypothesis was tested in a sample of 21 college-bound students and a sample of 15 students not planning to attend college. It was determined that 15 out of the 21 college-bound students, but only 5 out of the 15 noncollege-bound students, were marijuana smokers (see Chapter 10). Thus, we have the 2 × 2 problem in Table 11.4.

The relationship between college orientation and marijuana

TABLE 11.4
Marijuana smoking among college-oriented and noncollege-oriented students: data from Table 10.3

	Smokers	Nonsmokers	
College-oriented	15	6	21
Noncollege-oriented	5	10	15
	20	16	$N = 36$

smoking was tested by applying the 2×2 chi-square computational formula as follows:

$$\chi^2 = \frac{36[(15)(10) - (5)(6)]^2}{(15 + 5)(6 + 10)(15 + 6)(5 + 10)}$$

$$= \frac{36(150 - 30)^2}{(20)(16)(21)(15)}$$

$$= 5.14$$

Having calculated a chi-square value of 5.14, we can obtain the *phi coefficient* (ϕ), which is a measure of the degree of association for 2×2 tables. By formula,

$$\phi = \sqrt{\frac{\chi^2}{N}}$$

where

ϕ = the phi coefficient
χ^2 = the calculated chi-square value
N = total number of cases

Applying the foregoing formula to the problem at hand,

$$\phi = \sqrt{\frac{5.14}{36}}$$

$$= \sqrt{.14}$$

$$= .37$$

Our obtained phi coefficient of .37 indicates the presence of a moderate correlation between college orientation and marijuana smoking.

Testing the Significance of Phi

Fortunately, the phi coefficient can be easily tested by means of chi square, whose value has already been determined, and Table E in the back of the text:

obtained χ^2 = 5.14
table χ^2 = 3.84
df = 1
P = .05

Since our calculated chi-square value of 5.14 is greater than the required table value, we reject the null hypothesis that $\phi = 0$ and accept the research hypothesis that political orientation and marijuana smoking are associated in the population.

Requirements for the Use of the Phi Coefficient

In order to employ the phi coefficient as a measure of association between X and Y variables, we must consider the following requirements:

1. Nominal data—only frequency data are required.
2. A 2×2 table—the data must be capable of being cast in the form of a 2×2 table (2 rows by 2 columns). It is inappro-

priate to apply the phi coefficient to tables larger than 2×2, in which several groups or categories are being compared.

3. Random sampling—in order to test the significance of the phi coefficient, sample members must have been drawn on a random basis from a larger population.

CORRELATION COEFFICIENTS FOR NOMINAL DATA IN LARGER THAN 2×2 TABLES

Until this point, we have considered the correlation coefficient for nominal data arranged in a 2×2 table. As we have seen in Chapter 10, there are times when we have nominal data, but are comparing several groups or categories. To illustrate, let us reconsider the null hypothesis that

> *the relative frequency of permissive, moderate, and authoritarian child-rearing methods is the same for liberals, moderates, and conservatives.*

In Chapter 10, this hypothesis was tested with the data in the 3×3 table, Table 11.5.

The relationship between child-rearing method and political orientation was tested by applying the chi-square formula as follows:

$$\chi^2 = \frac{(7 - 10.79)^2}{10.79} + \frac{(10 - 10.07)^2}{10.07} + \frac{(15 - 11.14)^2}{11.14}$$
$$+ \frac{(9 - 10.11)^2}{10.11} + \frac{(10 - 9.44)^2}{9.44} + \frac{(11 - 10.45)^2}{10.45}$$
$$+ \frac{(14 - 9.10)^2}{9.10} + \frac{(8 - 8.49)^2}{8.49} + \frac{(5 - 9.40)^2}{9.40}$$
$$= 7.58$$

TABLE 11.5
Child rearing by political orientation: data from Table 10.4

	Conservative	Moderate	Liberal	
Permissive	7	9	14	30
Moderate	10	10	8	28
Authoritarian	15	11	5	31
	32	30	27	$N = 89$

In the present context, we seek to determine the correlation or degree of association between political orientation (X) and child-rearing method (Y). In a table larger than 2×2, this can be done by a simple extension of the chi-square test, which is referred to as the *contingency coefficient* (C). The value of C can be found by the formula

$$C = \sqrt{\frac{\chi^2}{N + \chi^2}}$$

where

χ^2 = the calculated chi-square value
N = the total number of cases
C = the contingency coefficient

In testing the degree of association between political orientation and child-rearing method,

$$C = \sqrt{\frac{7.58}{89 + 7.58}}$$
$$= \sqrt{\frac{7.58}{96.58}}$$
$$= \sqrt{.08}$$
$$= .28$$

Our obtained contingency coefficient of .28 indicates that the correlation between political orientation and child-rearing can be regarded as a rather weak one. Political orientation and child-rearing method are related, but many exceptions can be found.

Testing the Significance of the Contingency Coefficient

Just as in the case of the phi coefficient, whether the contingency coefficient is statistically significant can be easily determined from the size of the obtained chi-square value. In the present example, we find that the relationship between political orientation and child rearing is nonsignificant and, therefore, confined to the members of our samples. This is true since the calculated chi square of 7.58 is smaller than the required table value:

obtained $\chi^2 = 7.58$
table $\chi^2 = 9.49$
df $= 4$
$P = .05$

Requirements for the Use of the Contingency Coefficient

To appropriately apply the contingency coefficient, we must be aware of the following requirements:

1. Nominal data—only frequency data are required. These data may be cast in the form of a 2×2 table or larger.
2. Random sampling—for purposes of testing the significance

of the contingency coefficient, all sample members must have been taken at random from a larger population.

An Alternative to the Contingency Coefficient

Despite its great popularity among social researchers, the contingency coefficient has an important disadvantage: the number of rows and columns in a chi-square table will influence the maximum size taken by C. That is, the value of the contingency coefficient will not always vary between 0 and 1.0 (although it will never exceed 1.0). Under certain conditions, the maximum value of C may be .94; at other times, the maximum value of C may be .89; and so on.

To avoid this disadvantage of C, we may decide to employ another correlation coefficient, which expresses the degree of association between nominal-level variables in a larger than 2×2 table. Known as Cramér's V, this coefficient does not depend on the size of the χ^2 table and has the same requirements as the contingency coefficient. By formula,

$$V = \sqrt{\frac{\chi^2}{N(k - 1)}}$$

where

V = Cramér's V
N = the total number of cases
k = the number of rows *or* columns, whichever is smaller (if the number of rows equals the number of columns as in a 3×3, 4×4, or 5×5 table, either number can be used for k).

Returning to the relationship between political orientation and child rearing as shown in Table 11.5 (a 3×3 table),

$$V = \sqrt{\frac{7.58}{89(3 - 1)}}$$

$$= \sqrt{\frac{7.58}{89(2)}}$$

$$= \sqrt{\frac{7.58}{178}}$$

$$= \sqrt{.04}$$

$$= .20$$

Result: We find a Cramér's V correlation coefficient equal to .20, indicating a weak relationship between political orientation and child-rearing practices.

LAMBDA

As indicated in our previous discussion of Goodman's and Kruskal's Gamma, correlation can be regarded as the degree to which the values of one variable are predictable from knowledge of the values of the other variable.

Symbolized by the lower-case Greek letter λ, the *lambda* coefficient (also known as Guttman's coefficient of predictability) is a *proportionate reduction in error measure*—an index of how much we are able to reduce the error in predicting values of one variable from values of another. This is really another way of measuring to what degree we can improve the accuracy of our prediction. For example, if λ = .60, we have reduced the error of our prediction about values of the dependent variable by 60 percent; if λ = .25, we have reduced the error of our prediction by only 25 percent.

Like contingency coefficient and Cramér's *V*, lambda is a measure of association for comparing several groups or categories at the nominal level. Unlike these other coefficients which indicate mutual predictability, however, lambda is asymmetric—it measures the extent to which values of either variable (serving as the dependent variable) can be guessed or predicted from knowing values of the other variable (serving as the independent variable).

To illustrate, consider the relationship between an elementary school student's classroom placement and special needs label, either of which could be considered an independent variable for the other: A student's classroom placement (for example, in a "special class") might help determine whether or not he or she is labeled, or a student's label (for example, as mentally retarded or learning disabled) might influence his or her classroom placement. In reality, of course, there may be mutual predictability between classroom placement and label: Either variable probably has some capacity for predicting the other, but which do you think has the *greater* predictive ability, classroom placement or label?

Using lambda, we first treat classroom placement as the independent variable. We ask: To what extent can an individual's label be predicted by knowing whether he or she is placed in a special class, a regular class, or is tutored?

To find out, we arrange our data in the form of Table 11.6; in this example, a 3 × 3 table in which the categories of the independent variable (placement) are located in the columns and the categories of the dependent variable (labels) are located in the rows.

The value of λ can be found by the formula

$$\lambda = \frac{F_{iv} - M_{dv}}{N - M_{dv}}$$

where

λ = the lambda coefficient

F_{iv} = sum of the largest cell frequencies within each category of the independent variable

M_{dv} = the largest marginal total among categories of the dependent variable

N = the total number of cases

In testing the degree of association between placement (the inde-

TABLE 11.6
Placement and label

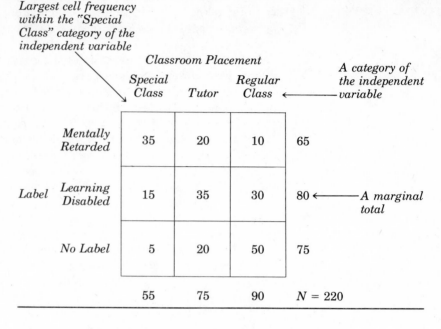

Largest cell frequency within the "Special Class" category of the independent variable

		Special Class	Tutor	Regular Class	
Label	*Mentally Retarded*	35	20	10	65
	Learning Disabled	15	35	30	80
	No Label	5	20	50	75
		55	75	90	$N = 220$

Classroom Placement — *A category of the independent variable*

A marginal total

pendent variable) and label (the dependent variable),

$$\lambda = \frac{(35 + 35 + 50) - 80}{220 - 80}$$

$$= \frac{120 - 80}{140}$$

$$= .29$$

The obtained lambda coefficient indicates that when classroom placement is treated as the independent variable, we reduce the error of our prediction (increase its accuracy) by 29 percent.

We next treat label as the independent variable. We now ask: To what degree can an individual's placement be predicted by knowing whether he or she has been labeled as mentally retarded or learning disabled, or has not been labeled?

Once again, we arrange our data in the form of a 3 × 3 table. This time, however, the categories of label are located in the columns and the categories of placement are located in the rows (see Table 11.7). In testing the degree of association between label (the independent variable) and placement (the dependent variable),

$$\lambda = \frac{(35 + 35 + 50) - 90}{220 - 90}$$

$$= \frac{120 - 90}{130}$$

$$= .23$$

TABLE 11.7
Label and placement

		Label			
		Mentally Retarded	Learning Disabled	No Label	
	Special Class	35	15	5	55
Classroom Placement	*Tutor*	20	35	20	75
	Regular Classroom	10	30	50	90
		65	80	75	N = 220

Using label as the independent variable, we are able to reduce the error of our prediction (increase its accuracy) by 23 percent. Conclusion: Classroom placement more accurately predicts label than label predicts classroom placement.

REQUIREMENTS FOR THE USE OF THE LAMBDA COEFFICIENT

To employ the lambda coefficient as a measure of association, the following requirements must be met:

1. Nominal data—only frequency data are required. Lambda can be applied to data cast in the form of a 2×2 table or larger.
2. Random sampling—in order to test the significance of the lambda coefficient (for example, with chi square), sample members must be drawn on a random basis from a specified population.

SUMMARY

In this chapter, we have been presented with correlation coefficients which numerically express the degree of association between X and Y variables. With the aid of the Pearson correlation coefficient (r), we can determine both the strength and direction of relationship between variables which have been measured at the interval level. We can also use Pearson r to predict values of one variable (Y) from knowledge of values of another variable (X).

There are several nonparametric alternatives to Pearson r. In order to determine the correlation between variables at the ordinal level of measurement, we can apply Spearman's rank-order correlation coefficient (r_s). To use this measure of correlation, both X and Y variables must be ranked or ordered. When a very large number of tied ranks occur, Goodman's and Kruskal's gamma coefficient (G) is a more effective alternative to the rank-order correlation coefficient.

By a simple extension of the chi-square test of significance, we can determine the degree of association between variables at the nominal level of measurement. For a 2×2 problem, we employ the phi coefficient (ϕ); for the larger than 2×2 problem, we use either the contingency coefficient or Cramér's V, or the asymmetric measure of association called lambda.

PROBLEMS

1. The six students below were questioned regarding (X) their attitudes toward the legalization of prostitution and (Y) their attitudes toward the legalization of marijuana. Compute a Pearson correlation coefficient for these data and determine whether the correlation is significant.

Student	X	Y
A	1	2
B	6	5
C	4	3
D	3	3
E	2	1
F	7	4

2. Compute a Pearson correlation coefficient for the following sets of scores and indicate whether the correlation is significant.

X	Y
2	5
1	4
5	3
4	1

3. Compute a Pearson correlation coefficient for the following set of scores and indicate whether the correlation is significant

X	Y
3	8
4	9
1	5
6	10
2	4

4. Compute a Pearson correlation coefficient for the following set of scores and indicate whether the correlation is significant.

X	Y
2	1
5	5
1	2
6	8
4	4

5. Compute a Pearson correlation coefficient for the following set of scores and indicate whether the correlation is significant.

X	Y
10	2
8	2
6	4
3	9
1	10
4	6
5	5

6. Employing the data in Problem 1 above, compute a regression equation to predict the value of Y (attitude toward legalization of marijuana) for the following values of X (attitude toward legalization of prostitution): (a) $X = 5$, (b) $X = 2$, (c) $X = 9$.

7. Employing the data in Problem 5 above, compute a regression equation to predict the value of Y for the following values of X: (a) $X = 10$; (b) $X = 2$.

8. The five students below were ranked in terms of when they completed an exam (1 = first to finish, 2 = second to finish, and so on) and were given an exam grade by the instructor. Test the null hypothesis of no relationship between (X) grade and (Y) length of time to complete the exam (that is, compute a rank-order correlation coefficient and indicate whether it is significant).

X	Y
53	1
91	2
70	3
85	4
91	5

9. The eight individuals below have been ranked on X and scored on Y. For these data, compute a rank-order correlation coefficient and indicate whether there is a significant relation between X and Y.

X	Y
1	32
2	28
3	45
4	60
5	45
6	60
7	53
8	55

10. The seven individuals below have been ranked on X and Y. For these data, compute a rank-order correlation coefficient and indicate whether there is a significant relation between X and Y.

X	Y
1	7
3	6
2	5
4	3
5	4
7	2
6	1

11. The five individuals below have been ranked from 1 to 5 on X and Y. For these data, compute a rank-order correlation coefficient and indicate whether there is a significant relation between X and Y.

X	Y
1	4
3	2
2	5
4	3
5	1

12. The five individuals below have been ranked from 1 to 5 on X and Y. For these data, compute a gamma coefficient and indicate whether there is a significant relation between X and Y.

X	Y
2	3
1	2
3	1
5	5
4	4

13. The 106 students below were ranked from high to low with respect to (X) their consumption of alcoholic beverages and (Y) their daily use of marijuana. For these data, compute a gamma coefficient to determine the degree of association between consumption of alcohol and use of marijuana, and indicate whether there is a significant relation between X and Y.

	Consumption of Alcohol		
	High	Medium	Low
Use of Marijuana	f	f	f
High	5	7	20
Medium	10	8	15
Low	15	6	10
		N = 106	

14. In Problem 2 of Chapter 10, $\chi^2 = 8.29$ for the relationship between attendance at lectures and final exam grades in statistics. Given the information that $N = 58$, compute a phi coefficient in order to determine the degree of association between these variables.

15. Given a 2×2 problem in which $N = 138$ and $\chi = 4.02$, compute a phi coefficient in order to determine the degree of association between the X and Y variables.

16. Given a 2×2 problem in which $N = 150$ and $\chi = 3.90$, compute a phi coefficient in order to determine the degree of association between the X and Y variables.

17. In order to determine the degree of association between X and Y for a 4×3 problem in which $N = 100$ and $\chi = 8.05$, compute (a) a contingency coefficient and (b) Cramér's V.

18. In Problem 5 of Chapter 10, it was ascertained that $N = 118$ and $\chi = 17.75$. Determine the degree of association between X and Y for this 4×2 problem (a) by computing a contingency coefficient, and (b) by Cramér's V.

19. In order to determine the degree of association between X and Y for a 3×3 problem in which $N = 138$ and $\chi = 10.04$, compute (a) a contingency coefficient and (b) Cramér's V.

20. The students below were asked to indicate (X) their college major and (Y) their political party affiliation. Using the lambda coefficient, determine by how much we can reduce the error in predicting values of one variable from values of the other. First, indicate to what extent students' college majors can be predicted by knowing their political party affiliations; then, indicate to what extent students' party affiliations can be predicted by knowing their college majors. Which variable, X or Y, has greater predictive ability?

College Major (X)	Political Party (Y)		
	Democrats f	Republicans f	Independents f
Liberal arts	45	31	36
Business	29	42	35
Education	22	10	13
			$N = 263$

TERMS TO REMEMBER

Variable
Correlation
   Strength
   Direction (positive versus negative)
   Curvilinear versus straight-line
Correlation coefficient
Scatter diagram
Regression line
Pearson r
Regression analysis
Regression equation
Rank-order correlation coefficient
Goodman's and Kruskal's gamma
Phi coefficient
Contingency coefficient
Cramér's V
Lambda

12 Applying Statistical Procedures to Research Problems

PART III OF the text contains a number of statistical procedures which can be applied to various problems in social research. Chapters 8, 9, and 10 introduced several different procedures for determining whether obtained sample differences are statistically significant or merely a product of sampling error. The procedures in Chapter 11 are designed to determine the degree of association, the correlation between two variables.

As noted throughout the text, each statistical procedure has a set of assumptions for its appropriate application. In selecting among procedures, any researcher must therefore consider a number of factors such as:

1. whether the researcher seeks to test for statistically significant differences, degree of association, or both;
2. whether the researcher has achieved the nominal, ordinal, or interval level of measurement of the variables being studied;
3. whether or not the variables being studied are normally distributed in the population from which they were drawn; and
4. whether the researcher is investigating independent samples or the same sample measured more than once.

Chapter 12 provides a number of hypothetical research situations in which the foregoing criteria are specified. The student is asked to choose the most appropriate statistical procedure for each research situation from among the following tests which were covered in Part III of the text:

1. *t* ratio
2. analysis of variance
3. chi square
4. median test
5. Kruskal-Wallis one-way analysis of variance
6. Friedman's two-way analysis of variance
7. Pearson *r*
8. Spearman's rank-order
9. Goodman's and Kruskal's gamma
10. phi
11. contingency coefficient
12. Cramér's *V*
13. Lambda

Table 12.1 (p. 244) locates each statistical procedure with respect to some of the important assumptions that must be considered for its appropriate application. Looking at the columns of the table, we face the first major decision related to the selection of a statistical procedure: Do we wish to determine whether or not a relationship exists? The tests of significance discussed in Chapters 8, 9, and 10 are designed to determine whether an obtained sample difference reflects a true population difference. Or, do we seek instead to establish the strength of the relationship between two variables? This is a question of correlation which can be addressed by means of the statistical procedures introduced in Chapter 11. Table 12.1 column subheadings indicate that a researcher who decides to employ a test of significance rather than a correlational procedure must also be aware of whether he is studying independent samples or the same sample measured more than once.

The rows of Table 12.1 direct our attention to the level at which our variables are measured. If we have achieved the interval level of measurement, we may well consider employing a parametric procedure such as *t*, *F*, or *r*. If, however, we have achieved either the nominal or ordinal level of measurement, the choice is limited to several nonparametric alternatives.

The solutions to the following research situations can be found at the end of the chapter.

RESEARCH SITUATIONS
Research Situation 1

A researcher conducted an experiment to determine the effect of a lecturer's age on student preferences to hear him lecture. In a regular classroom situation, 20 students were told that the administration wished to know their preferences regarding a forthcoming visiting lecturer series. In particular, they were asked to evaluate a professor who "might be visiting the campus." The professor was described to all students in the same way with one exception: One-half of the students were told the professor was 65 years old; one-half were told the professor was 25 years old. All students were then asked to indicate their willingness to attend the professor's lecture (higher scores indicate greater willingness). The following results were obtained:

X_1 (Scores of Students Told Professor Was 25 Years Old)	X_2 (Scores of Students Told Professor Was 65 Years Old)
65	78
38	42
52	77
71	50
69	65
72	70
55	55
78	51
56	33
80	59

Which statistical procedure would you apply to determine whether there is a significant difference between these groups of students with respect to their willingness to attend the lecture?

Research Situation 2   A researcher conducted an experiment to determine the effect of a lecturer's age on student preferences to hear him lecture. In a regular classroom situation, 30 students were told that the administration wished to know their preferences regarding a forthcoming visiting lecturer series. In particular, they were asked to evaluate a professor who "might be visiting the campus." The professor was described to all students in the same way with one exception: one-third of the students were told that the professor was 75 years old; one-third were told the professor was 50 years old; and one-third were told the professor was 25 years old. All students were then asked to indicate their willingness to attend the professor's lecture (higher scores indicate greater willingness). The following results were obtained:

X_1 (Scores of Students Told Professor Was 25 Years Old)	X_2 (Scores of Students Told Professor Was 50 Years Old)	X_3 (Scores of Students Told Professor Was 75 Years Old)
65	63	67
38	42	42
52	60	77
71	55	32
69	43	52
72	36	34
55	69	45
78	57	38
56	67	39
80	79	46

TABLE 12.1
Criteria for choosing an appropriate statistical procedure

Level of Measurement	Tests of Significance (Chapters 8, 9, 10)		Correlation (Chapter 11)
	Independent Samples	Same Sample Measured Twice	
Nominal	Chi square (nonparametric test for comparing two or more samples)		Phi coefficient (nonparametric for 2 × 2 tables) Contingency, Cramér's V, and lambda (nonparametric for larger than 2 × 2 tables)
Ordinal	Median test (nonparametric for comparing two samples) Kruskal-Wallis one-way analysis of variance (nonparametric for comparing three or more samples)	Friedman two-way analysis of variance (nonparametric for comparing same sample measured at least twice)	Spearman's rank-order (nonparametric) Goodman's and Kruskal's gamma (nonparametric for handling a large number of tied ranks)
Interval	t ratio (parametric for comparing two samples) Analysis of variance (parametric for comparing three or more samples)	t ratio (parametric for comparing the same sample measured twice)	Pearson r (parametric)

Which statistical procedure would you apply to determine whether there is a significant difference between these groups of students with respect to their willingness to attend the lecture?

Research Situation 3 To investigate the relationship between spelling and reading ability, a researcher gave spelling and reading examinations to a group of 20 students who had been selected at random from a large population of undergraduates. The following results were obtained (higher scores indicate greater ability):

Student	X (Spelling Score)	Y (Reading Score)
A	52	56
B	90	81
C	63	75
D	81	72
E	93	50
F	51	45
G	48	39
H	99	87
I	85	59
J	57	56
K	60	69
L	77	78
M	96	69
N	62	57
O	28	35
P	43	47
Q	88	73
R	72	76
S	75	63
T	69	79

Which statistical procedure would you apply to determine the degree of association between spelling and reading ability?

Research Situation 4 To investigate the validity of a particular reading test, researchers gave the reading test to a sample of 20 students whose ability to read had been previously ranked by their teacher. The test score and teacher's rank for each student are listed below:

Student	X (Reading Score)	Y (Teacher's Rank)
A	28	18
B	50	17
C	92	1
D	85	6
E	76	5

(Continued)

Student	X (Reading Score)	Y (Teacher's Rank)
F	69	10
G	42	11
H	53	12
I	80	3
J	91	2
K	73	4
L	74	9
M	14	20
N	29	19
O	86	7
P	73	8
Q	39	16
R	80	13
S	91	15
T	72	14

Which statistical procedure would you apply to determine the degree of association between reading scores and teacher's ranking?

Research Situation 5

To investigate regional differences in helpfulness toward strangers, a researcher dropped 400 keys (all of which had been stamped and tagged with a return address) around mailboxes in the northeastern, southern, midwestern, and western regions of the United States. The number of keys returned by region (as an indicator of helpfulness) is indicated below:

	Region			
	Northeast f	South f	Midwest f	West f
Returned	55	69	82	61
Not Returned	45	31	18	39
	100	100	100	100

Which statistical procedure would you apply to determine whether these regional differences are statistically significant?

Research Situation 6

To examine the relationship between authoritarianism and prejudice, a researcher administered measures of authoritarianism (the F scale) and prejudice (a checklist of negative adjectives to be assigned to black Americans) to a national sample of 950 adult Americans. The following results were obtained: Among 500 authoritarian respondents, 350 were "prejudiced" and 150 were "tolerant." Among 450 nonauthoritarian respondents, 125 were "prejudiced" and 325 were "tolerant."

Which statistical procedure would you apply to study the degree of association between authoritarianism and prejudice?

Research Situation 7   To investigate the relationship between year in school and grade-point average, researchers examined the academic records of 186 college students who were selected on a random basis from the undergraduate population of a certain university. The researchers obtained the following results:

	Year in School			
Grade-Point Average	1st f	2nd f	3d f	4th f
A− or better	6	5	7	10
B− to B+	10	16	19	18
C− to C+	23	20	15	7
C or worse	15	7	6	2
	54	48	47	37

(N = 186)

Which statistical procedure would you apply to determine the degree of association between grade-point average and year in school?

Research Situation 8   To investigate the influence of frustration on prejudice, 10 subjects were asked to assign negative adjectives such as lazy, dirty, and immoral to describe the members of a minority group (a measure of prejudice). All subjects described the minority group both before and after they had taken a series of lengthy and difficult examinations (the frustrating situation). The following results were obtained (higher scores represent greater prejudice):

Subject	X_1 (Prejudice Scores Before Taking the Frustrating Examinations)	X_2 (Prejudice Scores After Taking the Frustrating Examinations)
A	22	26
B	39	45
C	25	24
D	40	43
E	36	36
F	27	29
G	44	47
H	31	30
I	52	52
J	48	59

Which statistical procedure would you apply to determine whether there is a statistically significant difference in prejudice before and after the administration of the frustrating examinations?

Research Situation 9

To investigate the relationship between a respondent's actual occupational status and his subjective social class (that is, a respondent's own social class identification), 677 individuals were asked to indicate their occupation and the social class in which they belonged. Among 190 respondents with upper-status occupations (professional-technical-managerial), 56 identified themselves as upper class, 122 as middle class, and 12 as lower class; among 221 respondents with middle-status occupations (sales-clerical-skilled labor), 42 identified themselves as upper class, 163 as middle class, and 16 as lower class; among 266 with lower-status occupations (semi- and unskilled labor), 15 identified themselves as upper class, 202 as middle class, and 49 as lower class.
Which statistical procedure would you apply to determine the degree of association between occupational status and subjective social class?

Research Situation 10

To investigate the influence of college major on the starting salary of college graduates, researchers interviewed recent college graduates on their first jobs who had majored in engineering, liberal arts, or business administration. The results obtained for these 21 respondents are the following:

Starting Salaries

Engineering	Liberal Arts	Business Administration
$10,500	$ 7,000	$ 7,500
12,300	9,500	9,000
14,000	10,000	8,000
9,500	11,000	9,300
9,000	8,500	10,500
8,500	7,500	10,000
7,500	7,000	7,000

Which statistical procedure would you apply to determine whether there is a significant difference between these groups of respondents with respect to their starting salaries?

Research Situation 11

To investigate the influence of college major on the starting salary of college graduates, researchers interviewed recent college graduates on their first jobs who had majored in either liberal arts or business. The results obtained for these 16 respondents are the following:

Starting Salaries

Liberal Arts	Business
$ 7,000	$ 7,500
9,500	9,000
10,000	8,000
11,000	9,300
8,500	10,500
7,500	10,000
7,000	7,000
	8,000
	9,300

Which statistical procedure would you apply to determine whether there is a significant difference between liberal arts majors and business majors with respect to starting salaries?

Research Situation 12

A researcher conducted an experiment to determine the effect of a lecturer's age on student willingness to hear him lecture. In a regular classroom situation, 130 students were told that the administration wished to know their preferences regarding a forthcoming visiting lecturer series. In particular, they were asked to evaluate a professor who "might be visiting the campus." The professor was described to all students in the same way with one exception: One-half of the students were told the professor was 65 years old; one-half were told the professor was 25 years old. All students were then asked to indicate their willingness to attend the professor's lecture with the following results: Among those students told that the professor was 65, 22 expressed their willingness to attend his lecture and 43 expressed their unwillingness; among the students told that the professor was 25, 38 expressed their willingness to attend his lecture and 27 expressed their unwillingness.

Which statistical procedure would you apply to determine whether there is a significant difference between these groups of students with respect to their willingness to attend the professor's lecture?

Research Situation 13

To study the influence of teacher expectancy on student performance, achievement tests were given to a class of 15 third grade students both before and six weeks after their teacher was informed that all of them were "gifted." The following results were obtained (higher scores represent greater achievement):

Student	X_1 (Achievement Scores Before Teacher Expectancy)	X_2 (Achievement Scores After Teacher Expectancy)
A	98	100
B	112	113
C	101	101
D	124	125
E	92	91
F	143	145
G	103	105
H	110	115
I	115	119
J	98	99
K	117	119
L	93	99
M	108	105
N	102	103
O	136	140

Which statistical procedure would you apply to determine whether there is a statistically significant difference in achievement before and after the introduction of the teacher expectancy?

Research Situation 14

To investigate the relationship between neighborliness and voluntary association membership, a researcher questioned 500 public housing tenants concerning the amount of time they spend with other tenants (high/medium/low neighborliness) and the number of clubs and organizations in which they participate (high/medium/low voluntary association membership). Questionnaire results were as follows: Among 150 tenants who were "high" in regard to neighborliness, 60 were "high," 50 were "medium," and 40 were "low" with respect to voluntary association membership; among 180 tenants who were "medium" in regard to neighborliness, 45 were "high," 85 were "medium," and 50 were "low" with respect to voluntary association membership; and among 170 tenants who were "low" in regard to neighborliness, 40 were "high," 50 were "medium," and 80 were "low" with respect to voluntary association membership.

Which statistical procedure would you apply to determine the degree of association between neighborliness and voluntary association membership?

Research Situation 15

To determine the effect of special needs labels on teachers' judgments of their students' abilities, a researcher asked a sample of 40 elementary school teachers to evaluate the academic potential of an 11-year-old boy who was entering the sixth grade. All teachers were given "the report of a school psychologist" who described the boy in exactly the same way—with one exception: One fourth were told that the child was "emotionally disturbed," one fourth were told he was "mentally retarded," one fourth were told he was "learning disabled," and one fourth were given no special needs label. Immediately after reading about the boy, all teachers were asked to indicate on the numeric rating scale how successful they felt his academic progress would be (higher scores indicate greater optimism). The following scores were obtained:

X_1 (Scores for Emotionally Disturbed Label)	X_2 (Scores for Mentally Retarded Label)	X_3 (Scores for Learning Disabled Label)	X_4 (Scores for No Label)
23	29	25	42
29	42	46	51
41	32	53	31
36	25	56	37
37	37	44	40
56	48	41	43
45	42	38	55
39	28	32	52
28	30	50	42
32	32	43	24

Which statistical procedure would you apply to determine whether there is a significant difference between these groups of teachers with respect to their evaluations?

Research Situation 16

To study the relationship between income level and size of subordinate minority population, researchers determined the median income level for each of the following metropolitan areas of the United States which were also ranked with respect to the size of their black populations:

Metropolitan Area	X (Median Income)	Y (Rank with Respect to Size of Black Population)
A	8,500	2
B	12.400	6
C	7,900	1
D	10,200	3
E	15,600	7
F	14,100	5
G	9,300	4

Which statistical procedure would you apply to determine the degree of association between median income and size of black population?

Research Situation 17

To study the gossip of male and female college students, a researcher unobtrusively sat in the campus center and listened to students' conversations. The researcher was able to categorize the tone of students' gossip as negative (unfavorable statements about a third person), positive (favorable statements about a third person), or neutral. The following results were obtained: For 125 instances of gossip in the conversations of female students, 40 were positive, 36 were negative, and 49 were neutral. For 110 instances of gossip in the conversations of male students, 36 were positive, 32 were negative, and 42 were neutral.

Which statistical procedure would you apply to determine whether there is a significant difference between male and female college students with respect to the tone of their gossip?

Research Situation 18

To investigate the relationship between a respondent's political party affiliation and residential area, 328 individuals were asked to indicate their party preference (Republican, Democratic, or Independent) and the area in which their residence was located (urban, suburban, or rural). Among 124 Republicans, 25 lived in an urban area, 53 in the suburbs, and 46 in a rural area; among 142 Democrats, 61 lived in an urban area, 48 lived in the suburbs, and 33 lived in a rural area; among 62 Independents, 12 lived in an urban area, 26 lived in the suburbs, and 24 lived in a rural area.

Which statistical procedure would you apply to determine the degree of association between party affiliation and residential area?

RESEARCH SOLUTIONS
Solution to Research Situation 1
(t Ratio or Median Test)

Research situation 1 represents a comparison between the scores of two independent samples of students. The *t* ratio (Chapter 8) is employed in order to make comparisons between two means when interval data have been obtained. The median test (Chapter 10) is a nonparametric alternative which can be applied when we suspect that the scores are not normally distributed in the population or that the interval level of measurement has not been achieved.

Solution to Research Situation 2
(Analysis of Variance or Kruskal-Wallis One-Way Analysis of Variance)

Research situation 2 represents a comparison of the scores of three independent samples of students. The *F* ratio (analysis of variance—Chapter 9) is employed in order to make comparisons between three or more independent means when interval data have been obtained. Kruskal-Wallis one-way analysis of variance (Chapter 10) is a nonparametric alternative which can be applied when we have reason to suspect that the scores are not normally distributed in the population or when the interval level of measurement has not been achieved.

Solution to Research Situation 3 (*Pearson* r)

Research situation 3 is a correlation problem, since it asks for the degree of association between X (spelling ability) and Y (reading ability). Pearson r (Chapter 11) can be employed to detect a straight-line correlation between X and Y variables when both of these variables have been measured at the interval level. If X (spelling ability) and Y (reading ability) are not normally distributed in the population, we should consider applying a nonparametric alternative such as Spearman's rank-order correlation coefficient (Chapter 11).

Solution to Research Situation 4 (*Spearman's Rank-Order*)

Research situation 4 is a correlation problem, asking for the degree of association between X (reading scores) and Y (teacher's evaluation of reading ability). Spearman's rank-order correlation coefficient (Chapter 11) can be employed to detect a straight-line relationship between X and Y variables, when both of these variables have been ordered or ranked. Pearson r cannot be employed, since it requires interval level measurement of X and Y. In the present case, reading scores (X) must be ranked from 1 to 20 before rank-order is applied.

Solution to Research Situation 5 (*Chi Square*)

Research situation 5 represents a comparison between the frequencies (returned versus not returned) found in four groups (northeast, south, midwest, and west). The chi-square test of significance (Chapter 10) is used to make comparisons between two or more samples. Only nominal data are required. Present results can be cast in the form of a 2×4 table, representing 2 rows and 4 columns. Notice that the degree of association between return rate (X) and region (Y) can be measured by means of the contingency coefficient (C) or Cramér's V (Chapter 11).

Solution to Research Situation 6 (*Phi Coefficient*)

Research situation 6 is a correlation problem, which asks for the degree of association between X (authoritarianism) and Y (prejudice). The phi coefficient (Chapter 11) is a measure of association which can be employed when frequency or nominal data can be cast in the form of a 2×2 table (2 rows by 2 columns). In the present problem, such a table would take the following form:

Level of Prejudice	Level of Authoritarianism		
	Authoritarian	Nonauthoritarian	
Prejudiced	350	120	
Tolerant	150	325	$N = 950$

Solution to Research Situation 7
(Goodman's and Kruskal's Gamma)

Research situation 7 is a correlation problem, which asks for the degree of association between X (grade-point average) and Y (year in school). Goodman's and Kruskal's gamma coefficient (Chapter 11) is employed to detect a straight-line relationship between X and Y, when both variables have been ranked and a large number of ties has occurred. In the present problem, grade-point average has been ranked from A to D or worse and year in school has been ranked from 1st to 4th. Both of these crude ordinal measures have generated numerous tied ranks (for example, 54 students were in their first year of school; 48 students were in their second year, and so on). The contingency coefficient (C) or Cramér's V (Chapter 11) represents an alternative to gamma which assumes only nominal-level data.

Solution to Research Situation 8
(t Ratio or Two-Way Analysis of Variance by Ranks)

Research situation 8 represents a before-after comparison of a single sample measured at two different points in time. The t ratio (Chapter 8) can be employed to compare two means from a single sample arranged in a before-after panel design. Friedman's two-way analysis of variance (Chapter 10) is a nonparametric alternative which can be applied to the before-after situation when we have reason to suspect that the scores are not normally distributed in the population or when we have not achieved the interval level of measurement.

Solution to Research Situation 9
(Goodman's and Kruskal's Gamma)

Research situation 9 is a correlation problem, which asks for the degree of association between X (occupational status) and Y (subjective social class). Gamma (Chapter 11) is especially well-suited to the problem of detecting a straight-line relationship between X and Y, when both variables can be ranked and a large number of ties has occurred. In the present situation, occupational status and subjective social class have been ordered from "upper" to "middle" to "lower," generating a very large number of tied ranks (for example, 221 respondents had middle-status occupations). In order to obtain the gamma coefficient, the data must be rearranged in the form of a frequency table as follows:

Subjective Social Class (Y)	Occupational Status (X)		
	Upper *f*	Middle *f*	Lower *f*
Upper	56	42	15
Middle	122	163	202
Lower	12	16	49
	190	221	266

The contingency coefficient (C) and Cramér's V are alternatives to gamma which assume only nominal data.

Solution to Research Situation 10
(Analysis of Variance or Kruskal-Wallis One-Way Analysis of Variance)

Research situation 10 represents a comparison of the scores of three independent samples of respondents. The F ratio (Chapter 9) is used to make comparisons between three or more independent means when interval data have been obtained. Kruskal-Wallis one-way analysis of variance (Chapter 10) is a nonparametric alternative which can be employed when we suspect that the scores may not be normally distributed in the population or when the interval level of measurement has not been achieved.

Solution to Research Situation 11
(t Ratio or Median Test)

Research situation 11 represents a comparison between the scores of two independent samples of respondents. The t ratio (Chapter 8) is employed in order to compare two means when interval data have been obtained. The median test (Chapter 10) is a nonparametric alternative which can be applied when we cannot assume that the scores are normally distributed in the population or when the interval level of measurement has not been achieved.

Solution to Research Situation 12
(Chi Square)

Research situation 12 represents a comparison of the frequencies (willingness versus unwillingness) in two groups of students (those told the professor was 65 versus those told the professor was 25). The chi-square test of significance (Chapter 10) is used to make comparisons between two or more samples, when nominal or frequency data have been obtained. Present results can be cast in the form of the following 2 × 2 table, representing 2 rows and 2 columns:

	Experimental Condition		
Willingness to Attend	Students Told Professor Was 65 f	Students Told Professor Was 25 f	
Willing	22	38	
Unwilling	43	27	$N = 130$

Solution to Research Situation 13
(t Ratio or Two-Way Analysis of Variance by Ranks)

Research situation 13 represents a before-after comparison of a single sample measured at two different points in time. The t ratio (Chapter 8) can be used to compare two means from a single sample arranged in a before-after design. Friedman's two-way analysis of variance (Chapter 10) is a nonparametric alternative which can be applied to the before-after design if we suspect that the scores in the population do not take the shape of the normal distribution or if we have not achieved the interval level of measurement.

Solution to Research Situation 14
(Goodman's and Kruskal's Gamma)

Research situation 14 is a correlation problem which asks for the degree of association between X (neighborliness) and Y (voluntary association membership). Goodman's and Kruskal's gamma coefficient (Chapter 11) is applied in order to detect a straight-line relationship between X and Y, when both variables have been ranked and a large number of ties has occurred. In the present situation, both neighborliness and voluntary association participation have been ranked from high to low. Both of these crude ordinal measures have created numerous tied ranks (for example, 150 tenants were "high" in regard to neighborliness).

	Neighborliness (X)		
Voluntary Association Participation (Y)	Low *f*	Medium *f*	High *f*
Low	80	50	40
Medium	50	85	50
High	40	45	60
	170	180	150

The contingency coefficient (C) or Cramér's V (Chapter 11) represents an alternative to gamma which assumes only nominal-level data.

Solution to Research Situation 15
(Analysis of Variance or Kruskal-Wallis One-Way Analysis of Variance)

Research situation 15 represents a comparison of the scores of four independent samples of respondents. The F ratio (Chapter 9) is employed to make comparisons between three or more independent means when interval data have been achieved. Kruskal-Wallis one-way analysis of variance (Chapter 10) is a nonparametric alternative to be employed when we suspect that the scores may not be normally distributed in the population or when the interval level of measurement has not been achieved.

Solution to Research Situation 16
(Spearman's Rank-Order)

Research situation 16 is a correlation problem, asking for the degree of association between X (Median Income) and Y (Size of Black Population). Spearman's rank-order correlation coefficient (Chapter 11) can be applied to detect a straight-line relationship between X and Y variables, when both have been ranked or ordered. Pearson r cannot be employed because it requires interval data on X and Y. In the present case, median income (X) must be ranked from 1 to 7 before rank-order is applied.

Solution to Research Situation 17
(Chi Square)

Research situation 17 represents a comparison of the frequencies (negative, positive, and neutral tone) in female and male college students. The chi-square test of significance (Chapter 10) is employed to make comparisons between two or more samples, when nominal data have been obtained. Present results can be cast in the form of the following 2 × 3 table, representing 2 columns and 3 rows:

	Sex of Students	
Tone of Gossip	Female Conversations f	Male Conversations f
Negative	36	32
Positive	40	36
Neutral	49	42

Solution to Research Situation 18
(Lambda, Contingency Coefficient, or Cramér's V)

Research situation 18 is a correlation problem which asks for the degree of association between two variables measured at the nominal level: political party affiliation and residential area. Like contingency coefficient and Cramér's V, lambda (Chapter 11) is a measure of association for comparing several groups or categories at the nominal level. Unlike contingency and Cramér's V, lambda indicates the extent to which values of either variable can be predicted from knowing values of the other variable. To obtain lambda, contingency coefficient, or Cramér's V, the data must be arranged in the form of a frequency table as follows:

	Political Party Affiliation (X)		
Residential Area (Y)	Republican f	Democrat f	Independent f
Urban	25	61	12
Suburban	53	48	26
Rural	46	33	24

APPENDIXES

Appendix A
A Review of Some Fundamentals of Mathematics

FOR STUDENTS OF statistics who need to review some of the fundamentals of algebra and arithmetic, this appendix covers the problems of working with decimals, negative numbers, and square roots. Other problems of mathematics have been discussed at appropriate places throughout the text. For instance, Chapter 1 identifies, defines, and compares the three levels of measurement; Chapter 2 discusses percentages, proportions, ratios, and rates; and Chapter 4 explains summation (Σ).

WORKING WITH DECIMALS

When adding and subtracting decimals, be sure to place decimal points of the numbers directly below one another. For example, to add 3210.76, 2.541, and 98.3,

$$
\begin{array}{r}
3210.76 \\
2.541 \\
\underline{98.3} \\
3311.601
\end{array}
$$

To subtract 34.1 from 876.62,

$$
\begin{array}{r}
876.62 \\
\underline{-34.1} \\
842.52
\end{array}
$$

When multiplying decimals, be sure that your answer contains the same number of decimal places as both multiplicand and multiplier combined. For example,

Multiplicand →	63.41	2.6	.0003	.5
Multiplier →	×.05	×1.4	×.03	×.5
Product →	3.1705	3.64	.000009	.25

Before dividing, always eliminate decimals from the divisor by moving the decimal point as many places to the right as needed in order to make the divisor a whole number. Make a corresponding change of the same number of places for decimals in the dividend (that is, if you move the decimal two places in the divisor, then you must move it two places in the dividend). This procedure will indicate the number of decimal places in your answer.

$$\frac{2.44}{.02} = 122 \qquad \text{divisor} \longrightarrow .02\overline{)2.44} \longleftarrow \text{dividend, quotient}$$

$$\frac{2.2}{.4} = 2.2 \qquad .4\overline{).8\,8}$$

$$\frac{10.10}{10} = 1.01 \qquad 10\overline{)10.10}$$

$$\frac{1010}{.10} = 10100 \qquad .10\overline{)1010.00}$$

Arithmetic operations frequently yield answers in decimal form; for instance, 2.034, 24.7, 86.001, and so on. The question arises as to how many decimal places we should have in our answers. A simple rule to follow is to carry every operation to three more decimal places and round off to two more decimal places than found in the original set of numbers.

To illustrate, if data are derived from an original set of whole numbers (for instance, 12, 9, 49, or 15), we would carry out operations to three decimal places (to thousandths) and express our answer to the nearest hundredth. For example,

3.889 = 3.89
1.224 = 1.22
7.761 = 7.76

Rounding to the nearest decimal place is generally carried out as follows: Drop the last digit if it is less than 5 (in the examples below, the last digit is the thousandth digit):

less than 5

26.234 = 26.23
14.891 = 14.89
 1.012 = 1.01

Add one to the preceding digit if the last digit is 5 or more (in the examples below, the preceding digit is the hundredth digit):

5 or more

26.236 = 26.24
14.899 = 14.90
 1.015 = 1.02

The following have been rounded to the nearest whole number:

3.1 = 3
3.5 = 4
4.5 = 5
4.8 = 5

The following have been rounded to the nearest tenth:

3.11 = 3.1
3.55 = 3.6
4.45 = 4.5
4.17 = 4.2

The following have been rounded to the nearest hundredth:

3.328 = 3.33
4.823 = 4.82
3.065 = 3.07
3.055 = 3.06

DEALING WITH NEGATIVE NUMBERS

When adding a series of negative numbers, make sure that you give a negative sign to the sum. For example,

$$
\begin{array}{rr}
-20 & -3 \\
-12 & -9 \\
\underline{-6} & \underline{-4} \\
-38 & -16
\end{array}
$$

To add a series containing both negative and positive numbers, first group all negatives and all positives separately; add each group and then subtract their sums (the remainder gets the sign of the larger number). For example,

-6	$+4$	-6	$+6$
$+4$	$\underline{+2}$	-1	$\underline{-10}$
$+2$	$+6$	$\underline{-3}$	-4
-1		-10	
$\underline{-3}$			
-4			

To subtract a negative number, you must first give it a positive sign and then follow the procedure for addition. The remainder gets the sign of the larger number. For example,

$$\begin{array}{r} 24 \\ -(-6) \\ \hline 30 \end{array}$$
-6 gets a positive sign and is therefore added to 24. Since the larger value is a positive number (24), the remainder (30) is a positive value.

$$\begin{array}{r} -6 \\ -(-24) \\ \hline 18 \end{array}$$
-24 gets a positive sign and is therefore subtracted. Since the larger value is a positive number (remember that you have changed the sign of -24), the remainder (18) is a positive value.

$$\begin{array}{r} -24 \\ -(-6) \\ \hline -18 \end{array}$$
-6 gets a positive sign and is therefore subtracted. Since the larger value is a negative number (-24), the remainder (-18) is a negative value.

When multiplying (or dividing) two numbers that have the same sign, always assign a positive sign to their product (or quotient). For example,

$(+8) \times (+5) = +40$ $+5\overline{)+40}^{\,+8}$ $-5\overline{)-40}^{\,+8}$
$(-8) \times (-5) = +40$

In the case of two numbers having different signs, assign a negative sign to their product (or quotient). For example,

$(-8) \times (+5) = -40$ $-5\overline{)+40}^{\,-8}$

FINDING SQUARE ROOTS WITH TABLE A

With the aid of Table A in the back of the book, you can easily find the square root ($\sqrt{n}$) of any whole number (n) from 1 to 1000.

To approximate the square root of decimal numbers as well as numbers over 1000, it may be useful to begin with the column of squares (n^2) in Table A. The square root of any number multiplied by itself must equal that number. As a result, n in Table A is actually the square root of n^2.

In order to take full advantage of column n^2 for purposes of approximating square roots, we must determine how many digits precede the decimal point in any square root value. A simple rule to follow is to pair up the digits that come before the decimal point in a number. The number of pairs equals the number of digits that should be included in the square root of the number. For example,

$$\sqrt{\underline{55}\underline{55}.} = \underline{7}\underline{4}.53 \quad (2 \text{ pairs} = 2 \text{ digits})$$
$$\sqrt{\underline{55}.55} = \underline{7}.45 \quad (1 \text{ pair} = 1 \text{ digit})$$

When a number contains an odd number of digits, the odd digit preceding the decimal point adds another digit to the square root of the number, just as if it were a complete pair. For example,

$$\sqrt{\underline{5}\underline{55}.5} = \underline{2}\underline{3}.57 \quad (1 \text{ pair} + 1 \text{ odd digit} = 2 \text{ digits})$$
$$\sqrt{\underline{5}.555} = \underline{2}.36 \quad (1 \text{ odd digit} = 1 \text{ digit})$$

To find the square root of any number less than 1, the following procedure may be carried out:

1. Round to the nearest hundredth

$$\sqrt{.328} = \sqrt{.33}$$
$$\sqrt{.823} = \sqrt{.82}$$
$$\sqrt{.0651} = \sqrt{.07}$$
$$\sqrt{.035} = \sqrt{.04}$$

2. Locate the square root of the corresponding whole number in Table A (to find the whole number, simply drop the decimal point)

$$\sqrt{33} = 5.74$$
$$\sqrt{82} = 9.06$$
$$\sqrt{7} = 2.65$$
$$\sqrt{4} = 2$$

3. Move the decimal point one place to the left and round off

$$\sqrt{.33} = .57$$
$$\sqrt{.82} = .91$$
$$\sqrt{.07} = .27$$
$$\sqrt{.04} = .2$$

Appendix B
Tables

TABLE A
Squares, Square Roots, and Reciprocals of the numbers from 1 to 1000

n	n^2	$\sqrt{n}$	$\dfrac{1}{n}$	$\dfrac{1}{\sqrt{n}}$
1	1	1.0000	1.000000	1.0000
2	4	1.4142	.500000	.7071
3	9	1.7321	.333333	.5774
4	16	2.0000	.250000	.5000
5	25	2.2361	.200000	.4472
6	36	2.4495	.166667	.4082
7	49	2.6458	.142857	.3780
8	64	2.8284	.125000	.3536
9	81	3.0000	.111111	.3333
10	100	3.1623	.100000	.3162
11	121	3.3166	.090909	.3015
12	144	3.4641	.083333	.2887
13	169	3.6056	.076923	.2774
14	196	3.7417	.071429	.2673
15	225	3.8730	.066667	.2582
16	256	4.0000	.062500	.2500
17	289	4.1231	.058824	.2425
18	324	4.2426	.055556	.2357
19	361	4.3589	.052632	.2294
20	400	4.4721	.050000	.2236
21	441	4.5826	.047619	.2182
22	484	4.6904	.045455	.2132
23	529	4.7958	.043478	.2085
24	576	4.8990	.041667	.2041
25	625	5.0000	.040000	.2000
26	676	5.0990	.038462	.1961
27	729	5.1962	.037037	.1925
28	784	5.2915	.035714	.1890
29	841	5.3852	.034483	.1857
30	900	5.4772	.033333	.1826
31	961	5.5678	.032258	.1796
32	1024	5.6569	.031250	.1768
33	1089	5.7446	.030303	.1741
34	1156	5.8310	.029412	.1715
35	1225	5.9161	.028571	.1690

n	n^2	$\sqrt{n}$	$\dfrac{1}{n}$	$\dfrac{1}{\sqrt{n}}$
36	1296	6.0000	.027778	.1667
37	1369	6.0828	.027027	.1644
38	1444	6.1644	.026316	.1622
39	1521	6.2450	.025641	.1601
40	1600	6.3246	.025000	.1581
41	1681	6.4031	.024390	.1562
42	1764	6.4807	.023810	.1543
43	1849	6.5574	.023256	.1525
44	1936	6.6332	.022727	.1508
45	2025	6.7082	.022222	.1491
46	2116	6.7823	.021739	.1474
47	2209	6.8557	.021277	.1459
48	2304	6.9282	.020833	.1443
49	2401	7.0000	.020408	.1429
50	2500	7.0711	.020000	.1414
51	2601	7.1414	.019608	.1400
52	2704	7.2111	.019231	.1387
53	2809	7.2801	.018868	.1374
54	2916	7.3485	.018519	.1361
55	3025	7.4162	.018182	.1348
56	3136	7.4833	.017857	.1336
57	3249	7.5498	.017544	.1325
58	3364	7.6158	.017241	.1313
59	3481	7.6811	.016949	.1302
60	3600	7.7460	.016667	.1291
61	3721	7.8102	.016393	.1280
62	3844	7.8740	.016129	.1270
63	3969	7.9373	.015873	.1260
64	4096	8.0000	.015625	.1250
65	4225	8.0623	.015385	.1240
66	4356	8.1240	.015152	.1231
67	4489	8.1854	.014925	.1222
68	4624	8.2462	.014706	.1213
69	4761	8.3066	.014493	.1204
70	4900	8.3666	.014286	.1195
71	5041	8.4261	.014085	.1187
72	5184	8.4853	.013889	.1179
73	5329	8.5440	.013699	.1170
74	5476	8.6023	.013514	.1162
75	5625	8.6603	.013333	.1155
76	5776	8.7178	.013158	.1147
77	5929	8.7750	.012987	.1140
78	6084	8.8318	.012821	.1132
79	6241	8.8882	.012658	.1125
80	6400	8.9443	.012500	.1118
81	6561	9.0000	.012346	.1111
82	6724	9.0554	.012195	.1104
83	6889	9.1104	.012048	.1098
84	7056	9.1652	.011905	.1091
85	7225	9.2195	.011765	.1085

Tables

n	n^2	$\sqrt{n}$	$\dfrac{1}{n}$	$\dfrac{1}{\sqrt{n}}$
86	7396	9.2736	.011628	.1078
87	7569	9.3274	.011494	.1072
88	7744	9.3808	.011364	.1066
89	7921	9.4340	.011236	.1060
90	8100	9.4868	.011111	.1054
91	8281	9.5394	.010989	.1048
92	8464	9.5917	.010870	.1043
93	8649	9.6437	.010753	.1037
94	8836	9.6954	.010638	.1031
95	9025	9.7468	.010526	.1026
96	9216	9.7980	.010417	.1021
97	9409	9.8489	.010309	.1015
98	9604	9.8995	.010204	.1010
99	9801	9.9499	.010101	.1005
100	10000	10.0000	.010000	.1000
101	10201	10.0499	.009901	.0995
102	10404	10.0995	.009804	.0990
103	10609	10.1489	.009709	.0985
104	10816	10.1980	.009615	.0981
105	11025	10.2470	.009524	.0976
106	11236	10.2956	.009434	.0971
107	11449	10.3441	.009346	.0967
108	11664	10.3923	.009259	.0962
109	11881	10.4403	.009174	.0958
110	12100	10.4881	.009091	.0953
111	12321	10.5357	.009009	.0949
112	12544	10.5830	.008929	.0945
113	12769	10.6301	.008850	.0941
114	12996	10.6771	.008772	.0937
115	13225	10.7238	.008696	.0933
116	13456	10.7703	.008621	.0928
117	13689	10.8167	.008547	.0925
118	13924	10.8628	.008475	.0921
119	14161	10.9087	.008403	.0917
120	14400	10.9545	.008333	.0913
121	14641	11.0000	.008264	.0909
122	14884	11.0454	.008197	.0905
123	15129	11.0905	.008130	.0902
124	15376	11.1355	.008065	.0898
125	15625	11.1803	.008000	.0894
126	15876	11.2250	.007937	.0891
127	16129	11.2694	.007874	.0887
128	16384	11.3137	.007813	.0884
129	16641	11.3578	.007752	.0880
130	16900	11.4018	.007692	.0877
131	17161	11.4455	.007634	.0874
132	17424	11.4891	.007576	.0870
133	17689	11.5326	.007519	.0867
134	17956	11.5758	.007463	.0864
135	18225	11.6190	.007407	.0861

TABLE A
(*Continued*)

n	n^2	$\sqrt{n}$	$\dfrac{1}{n}$	$\dfrac{1}{\sqrt{n}}$
136	18496	11.6619	.007353	.0857
137	18769	11.7047	.007299	.0854
138	19044	11.7473	.007246	.0851
139	19321	11.7898	.007194	.0848
140	19600	11.8322	.007143	.0845
141	19881	11.8743	.007092	.0842
142	20164	11.9164	.007042	.0839
143	20449	11.9583	.006993	.0836
144	20736	12.0000	.006944	.0833
145	21025	12.0416	.006897	.0830
146	21316	12.0830	.006849	.0828
147	21609	12.1244	.006803	.0825
148	21904	12.1655	.006757	.0822
149	22201	12.2066	.006711	.0819
150	22500	12.2474	.006667	.0816
151	22801	12.2882	.006623	.0814
152	23104	12.3288	.006579	.0811
153	23409	12.3693	.006536	.0808
154	23716	12.4097	.006494	.0806
155	24025	12.4499	.006452	.0803
156	24336	12.4900	.006410	.0801
157	24649	12.5300	.006369	.0798
158	24964	12.5698	.006329	.0796
159	25281	12.6095	.006289	.0793
160	25600	12.6491	.006250	.0791
161	25921	12.6886	.006211	.0788
162	26244	12.7279	.006173	.0786
163	26569	12.7671	.006135	.0783
164	26896	12.8062	.006098	.0781
165	27225	12.8452	.006061	.0778
166	27556	12.8841	.006024	.0776
167	27889	12.9228	.005988	.0774
168	28224	12.9615	.005952	.0772
169	28561	13.0000	.005917	.0769
170	28900	13.0384	.005882	.0767
171	29241	13.0767	.005848	.0765
172	29584	13.1149	.005814	.0762
173	29929	13.1529	.005780	.0760
174	30276	13.1909	.005747	.0758
175	30625	13.2288	.005714	.0756
176	30976	13.2665	.005682	.0754
177	31329	13.3041	.005650	.0752
178	31684	13.3417	.005618	.0750
179	32041	13.3791	.005587	.0747
180	32400	13.4164	.005556	.0745
181	32761	13.4536	.005525	.0743
182	33124	13.4907	.005495	.0741
183	33489	13.5277	.005464	.0739
184	33856	13.5647	.005435	.0737
185	34225	13.6015	.005405	.0735

TABLE A
(*Continued*)

n	n^2	$\sqrt{n}$	$\dfrac{1}{n}$	$\dfrac{1}{\sqrt{n}}$
186	34596	13.6382	.005376	.0733
187	34969	13.6748	.005348	.0731
188	35344	13.7113	.005319	.0729
189	35721	13.7477	.005291	.0727
190	36100	13.7840	.005263	.0725
191	36481	13.8203	.005236	.0724
192	36864	13.8564	.005208	.0722
193	37249	13.8924	.005181	.0720
194	37636	13.9284	.005155	.0718
195	38025	13.9642	.005128	.0716
196	38416	14.0000	.005102	.0714
197	38809	14.0357	.005076	.0712
198	39204	14.0712	.005051	.0711
199	39601	14.1067	.005025	.0709
200	40000	14.1421	.005000	.0707
201	40401	14.1774	.004975	.0705
202	40804	14.2127	.004950	.0704
203	41209	14.2478	.004926	.0702
204	41616	14.2829	.004902	.0700
205	42025	14.3178	.004878	.0698
206	42436	14.3527	.004854	.0697
207	42849	14.3875	.004831	.0695
208	43264	14.4222	.004808	.0693
209	43681	14.4568	.004785	.0692
210	44100	14.4914	.004762	.0690
211	44521	14.5258	.004739	.0688
212	44944	14.5602	.004717	.0687
213	45369	14.5945	.004695	.0685
214	45796	14.6287	.004673	.0684
215	46225	14.6629	.004651	.0682
216	46656	14.6969	.004630	.0680
217	47089	14.7309	.004608	.0679
218	47524	14.7648	.004587	.0677
219	47961	14.7986	.004566	.0676
220	48400	14.8324	.004545	.0674
221	48841	14.8661	.004525	.0673
222	49284	14.8997	.004505	.0671
223	49729	14.9332	.004484	.0670
224	50176	14.9666	.004464	.0668
225	50625	15.0000	.004444	.0667
226	51076	15.0333	.004425	.0665
227	51529	15.0665	.004405	.0664
228	51984	15.0997	.004386	.0662
229	52441	15.1327	.004367	.0661
230	52900	15.1658	.004348	.0659
231	53361	15.1987	.004329	.0658
232	53824	15.2315	.004310	.0657
233	54289	15.2643	.004292	.0655
234	54756	15.2971	.004274	.0654
235	55225	15.3297	.004255	.0652

TABLE A
(*Continued*)

n	n^2	$\sqrt{n}$	$\dfrac{1}{n}$	$\dfrac{1}{\sqrt{n}}$
236	55696	15.3623	.004237	.0651
237	56169	15.3948	.004219	.0650
238	56644	15.4272	.004202	.0648
239	57121	15.4596	.004184	.0647
240	57600	15.4919	.004167	.0645
241	58081	15.5242	.004149	.0644
242	58564	15.5563	.004132	.0643
243	59049	15.5885	.004115	.0642
244	59536	15.6205	.004098	.0640
245	60025	15.6525	.004082	.0639
246	60516	15.6844	.004065	.0638
247	61009	15.7162	.004049	.0636
248	61504	15.7480	.004032	.0635
249	62001	15.7797	.004016	.0634
250	62500	15.8114	.004000	.0632
251	63001	15.8430	.003984	.0631
252	63504	15.8745	.003968	.0630
253	64009	15.9060	.003953	.0629
254	64516	15.9374	.003937	.0627
255	65025	15.9687	.003922	.0626
256	65536	16.0000	.003906	.0625
257	66049	16.0312	.003891	.0624
258	66564	16.0624	.003876	.0623
259	67081	16.0935	.003861	.0621
260	67600	16.1245	.003846	.0620
261	68121	16.1555	.003831	.0619
262	68644	16.1864	.003817	.0618
263	69169	16.2173	.003802	.0617
264	69696	16.2481	.003788	.0615
265	70225	16.2788	.003774	.0614
266	70756	16.3095	.003759	.0613
267	71289	16.3401	.003745	.0612
268	71824	16.3707	.003731	.0611
269	72361	16.4012	.003717	.0610
270	72900	16.4317	.003704	.0609
271	73441	16.4621	.003690	.0607
272	73984	16.4924	.003676	.0606
273	74529	16.5227	.003663	.0605
274	75076	16.5529	.003650	.0604
275	75625	16.5831	.003636	.0603
276	76176	16.6132	.003623	.0602
277	76729	16.6433	.003610	.0601
278	77284	16.6733	.003597	.0600
279	77841	16.7033	.003584	.0599
280	78400	16.7332	.003571	.0598
281	78961	16.7631	.003559	.0597
282	79524	16.7929	.003546	.0595
283	80089	16.8226	.003534	.0594
284	80656	16.8523	.003521	.0593
285	81225	16.8819	.003509	.0592

n	n^2	$\sqrt{n}$	$\dfrac{1}{n}$	$\dfrac{1}{\sqrt{n}}$
286	81796	16.9115	.003497	.0591
287	82369	16.9411	.003484	.0590
288	82944	16.9706	.003472	.0589
289	83521	17.0000	.003460	.0588
290	84100	17.0294	.003448	.0587
291	84681	17.0587	.003436	.0586
292	85264	17.0880	.003425	.0585
293	85849	17.1172	.003413	.0584
294	86436	17.1464	.003401	.0583
295	87025	17.1756	.003390	.0582
296	87616	17.2047	.003378	.0581
297	88209	17.2337	.003367	.0580
298	88804	17.2627	.003356	.0579
299	89401	17.2916	.003344	.0578
300	90000	17.3205	.003333	.0577
301	90601	17.3494	.003322	.0576
302	91204	17.3781	.003311	.0575
303	91809	17.4069	.003300	.0574
304	92416	17.4356	.003289	.0574
305	93025	17.4642	.003279	.0573
306	93636	17.4929	.003268	.0572
307	94249	17.5214	.003257	.0571
308	94864	17.5499	.003247	.0570
309	95481	17.5784	.003236	.0569
310	96100	17.6068	.003226	.0568
311	96721	17.6352	.003215	.0567
312	97344	17.6635	.003205	.0566
313	97969	17.6918	.003195	.0565
314	98596	17.7200	.003185	.0564
315	99225	17.7482	.003175	.0563
316	99856	17.7764	.003165	.0563
317	100489	17.8045	.003155	.0562
318	101124	17.8326	.003145	.0561
319	101761	17.8606	.003135	.0560
320	102400	17.8885	.003125	.0559
321	103041	17.9165	.003115	.0558
322	103684	17.9444	.003106	.0557
323	104329	17.9722	.003096	.0556
324	104976	18.0000	.003086	.0556
325	105625	18.0278	.003077	.0555
326	106276	18.0555	.003067	.0554
327	106929	18.0831	.003058	.0553
328	107584	18.1108	.003049	.0552
329	108241	18.1384	.003040	.0551
330	108900	18.1659	.003030	.0550
331	109561	18.1934	.003021	.0550
332	110224	18.2209	.003012	.0549
333	110889	18.2483	.003003	.0548
334	111556	18.2757	.002994	.0547
335	112225	18.3030	.002985	.0546

TABLE A
(Continued)

n	n^2	$\sqrt{n}$	$\dfrac{1}{n}$	$\dfrac{1}{\sqrt{n}}$
336	112896	18.3303	.002976	.0546
337	113569	18.3576	.002967	.0545
338	114244	18.3848	.002959	.0544
339	114921	18.4120	.002950	.0543
340	115600	18.4391	.002941	.0542
341	116281	18.4662	.002933	.0542
342	116964	18.4932	.002924	.0541
343	117649	18.5203	.002915	.0540
344	118336	18.5472	.002907	.0539
345	119025	18.5742	.002899	.0538
346	119716	18.6011	.002890	.0538
347	120409	18.6279	.002882	.0537
348	121104	18.6548	.002874	.0536
349	121801	18.6815	.002865	.0535
350	122500	18.7083	.002857	.0535
351	123201	18.7350	.002849	.0534
352	123904	18.7617	.002841	.0533
353	124609	18.7883	.002833	.0532
354	125316	18.8149	.002825	.0531
355	126025	18.8414	.002817	.0531
356	126736	18.8680	.002809	.0530
357	127449	18.8944	.002801	.0529
358	128164	18.9209	.002793	.0529
359	128881	18.9473	.002786	.0528
360	129600	18.9737	.002778	.0527
361	130321	19.0000	.002770	.0526
362	131044	19.0263	.002762	.0526
363	131769	19.0526	.002755	.0525
364	132496	19.0788	.002747	.0524
365	133225	19.1050	.002740	.0523
366	133956	19.1311	.002732	.0523
367	134689	19.1572	.002725	.0522
368	135424	19.1833	.002717	.0521
369	136161	19.2094	.002710	.0521
370	136900	19.2354	.002703	.0520
371	137641	19.2614	.002695	.0519
372	138384	19.2873	.002688	.0518
373	139129	19.3132	.002681	.0518
374	139876	19.3391	.002674	.0517
375	140625	19.3649	.002667	.0516
376	141376	19.3907	.002660	.0516
377	142129	19.4165	.002653	.0515
378	142884	19.4422	.002646	.0514
379	143641	19.4679	.002639	.0514
380	144400	19.4936	.002632	.0513
381	145161	19.5192	.002625	.0512
382	145924	19.5448	.002618	.0512
383	146689	19.5704	.002611	.0511
384	147456	19.5959	.002604	.0510
385	148225	19.6214	.002597	.0510

TABLE A
(*Continued*)

n	n^2	$\sqrt{n}$	$\dfrac{1}{n}$	$\dfrac{1}{\sqrt{n}}$
386	148996	19.6469	.002591	.0509
387	149769	19.6723	.002584	.0508
388	150544	19.6977	.002577	.0508
389	151321	19.7231	.002571	.0507
390	152100	19.7484	.002564	.0506
391	152881	19.7737	.002558	.0506
392	153664	19.7990	.002551	.0505
393	154449	19.8242	.002545	.0504
394	155236	19.8494	.002538	.0504
395	156025	19.8746	.002532	.0503
396	156816	19.8997	.002525	.0503
397	157609	19.9249	.002519	.0502
398	158404	19.9499	.002513	.0501
399	159201	19.9750	.002506	.0501
400	160000	20.0000	.002500	.0500
401	160801	20.0250	.002494	.0499
402	161604	20.0499	.002488	.0499
403	162409	20.0749	.002481	.0498
404	163216	20.0998	.002475	.0498
405	164025	20.1246	.002469	.0497
406	164836	20.1494	.002463	.0496
407	165649	20.1742	.002457	.0496
408	166464	20.1990	.002451	.0495
409	167281	20.2237	.002445	.0494
410	168100	20.2485	.002439	.0494
411	168921	20.2731	.002433	.0493
412	169744	20.2978	.002427	.0493
413	170569	20.3224	.002421	.0492
414	171396	20.3470	.002415	.0491
415	172225	20.3715	.002410	.0491
416	173056	20.3961	.002404	.0490
417	173889	20.4206	.002398	.0490
418	174724	20.4450	.002392	.0489
419	175561	20.4695	.002387	.0489
420	176400	20.4939	.002381	.0488
421	177241	20.5183	.002375	.0487
422	178084	20.5426	.002370	.0487
423	178929	20.5670	.002364	.0486
424	179776	20.5913	.002358	.0486
425	180625	20.6155	.002353	.0485
426	181476	20.6398	.002347	.0485
427	182329	20.6640	.002342	.0484
428	183184	20.6882	.002336	.0483
429	184041	20.7123	.002331	.0483
430	184900	20.7364	.002326	.0482
431	185761	20.7605	.002320	.0482
432	186624	20.7846	.002315	.0481
433	187489	20.8087	.002309	.0481
434	188356	20.8327	.002304	.0480
435	189225	20.8567	.002299	.0479

n	n^2	$\sqrt{n}$	$\dfrac{1}{n}$	$\dfrac{1}{\sqrt{n}}$
436	190096	20.8806	.002294	.0479
437	190969	20.9045	.002288	.0478
438	191844	20.9284	.002283	.0478
439	192721	20.9523	.002278	.0477
440	193600	20.9762	.002273	.0477
441	194481	21.0000	.002268	.0476
442	195364	21.0238	.002262	.0476
443	196249	21.0476	.002257	.0475
444	197136	21.0713	.002252	.0475
445	198025	21.0950	.002247	.0474
446	198916	21.1187	.002242	.0474
447	199809	21.1424	.002237	.0473
448	200704	21.1660	.002232	.0472
449	201601	21.1896	.002227	.0472
450	202500	21.2132	.002222	.0471
451	203401	21.2368	.002217	.0471
452	204304	21.2603	.002212	.0470
453	205209	21.2838	.002208	.0470
454	206116	21.3073	.002203	.0469
455	207025	21.3307	.002198	.0469
456	207936	21.3542	.002193	.0468
457	208849	21.3776	.022188	.0468
458	209764	21.4009	.002183	.0467
459	210681	21.4243	.002179	.0467
460	211600	21.4476	.002174	.0466
461	212521	21.4709	.002169	.0466
462	213444	21.4942	.002165	.0465
463	214369	21.5174	.002160	.0465
464	215296	21.5407	.002155	.0464
465	216225	21.5639	.002151	.0464
466	217156	21.5870	.002146	.0463
467	218089	21.6102	.002141	.0463
468	219024	21.6333	.002137	.0462
469	219961	21.6564	.002132	.0462
470	220900	21.6795	.002128	.0461
471	221841	21.7025	.002123	.0461
472	222784	21.7256	.002119	.0460
473	223729	21.7486	.002114	.0460
474	224676	21.7715	.002110	.0459
475	225625	21.7945	.002105	.0459
476	226576	21.8174	.002101	.0458
477	227529	21.8403	.002096	.0458
478	228484	21.8632	.002092	.0457
479	229441	21.8861	.002088	.0457
480	230400	21.9089	.002083	.0456
481	231361	21.9317	.002079	.0456
482	232324	21.9545	.002075	.0455
483	233289	21.9773	.002070	.0455
484	234256	22.0000	.002066	.0455
485	235225	22.0227	.002062	.0454

n	n^2	$\sqrt{n}$	$\dfrac{1}{n}$	$\dfrac{1}{\sqrt{n}}$
486	236196	22.0454	.002058	.0454
487	237169	22.0681	.002053	.0453
488	238144	22.0907	.002049	.0453
489	239121	22.1133	.002045	.0452
490	240100	22.1359	.002041	.0452
491	241081	22.1585	.002037	.0451
492	242064	22.1811	.002033	.0451
493	243049	22.2036	.002028	.0450
494	244036	22.2261	.002024	.0450
495	245025	22.2486	.002020	.0449
496	246016	22.2711	.002016	.0448
497	247009	22.2935	.002012	.0449
498	248004	22.3159	.002008	.0449
499	249001	22.3383	.002004	.0448
500	250000	22.3607	.002000	.0447
501	251001	22.3830	.001996	.0447
502	252004	22.4054	.001992	.0446
503	253009	22.4277	.001988	.0446
504	254016	22.4499	.001984	.0445
505	255025	22.4722	.001980	.0445
506	256036	22.4944	.001976	.0445
507	257049	22.5167	.001972	.0444
508	258064	22.5389	.001969	.0444
509	259081	22.5610	.001965	.0443
510	260100	22.5832	.001961	.0443
511	261121	22.6053	.001957	.0442
512	262144	22.6274	.001953	.0442
513	263169	22.6495	.001949	.0442
514	264196	22.6716	.001946	.0441
515	265225	22.6936	.001942	.0441
516	266256	22.7156	.001938	.0440
517	267289	22.7376	.001934	.0440
518	268324	22.7596	.001931	.0439
519	269361	22.7816	.001927	.0439
520	270400	22.8035	.001923	.0439
521	271441	22.8254	.001919	.0438
522	272484	22.8473	.001916	.0438
523	273529	22.8692	.001912	.0437
524	274576	22.8910	.001908	.0437
525	275625	22.9129	.001905	.0436
526	276676	22.9347	.001901	.0436
527	277729	22.9565	.001898	.0436
528	278784	22.9783	.001894	.0435
529	279841	23.0000	.001890	.0435
530	280900	23.0217	.001887	.0434
531	281961	23.0434	.001883	.0434
532	283024	23.0651	.001880	.0434
533	284089	23.0868	.001876	.0433
534	285156	23.1084	.001873	.0433
535	286225	23.1301	.001869	.0432

TABLE A
(*Continued*)

n	n^2	$\sqrt{n}$	$\dfrac{1}{n}$	$\dfrac{1}{\sqrt{n}}$
536	287296	23.1517	.001866	.0432
537	288369	23.1733	.001862	.0432
538	289444	23.1948	.001859	.0431
539	290521	23.2164	.001855	.0431
540	291600	23.2379	.001852	.0430
541	292681	23.2594	.001848	.0430
542	293764	23.2809	.001845	.0430
543	294849	23.3024	.001842	.0429
544	295936	23.3238	.001838	.0429
545	297025	23.3452	.001835	.0428
546	298116	23.3666	.001832	.0428
547	299209	23.3880	.001828	.0428
548	300304	23.4094	.001825	.0427
549	301401	23.4307	.001821	.0427
550	302500	23.4521	.001818	.0426
551	303601	23.4734	.001815	.0426
552	304704	23.4947	.001812	.0426
553	305809	23.5160	.001808	.0425
554	306916	23.5372	.001805	.0425
555	308025	23.5584	.001802	.0424
556	309136	23.5797	.001799	.0424
557	310249	23.6008	.001795	.0424
558	311364	23.6220	.001792	.0423
559	312481	23.6432	.001789	.0423
560	313600	23.6643	.001786	.0423
561	314721	23.6854	.001783	.0422
562	315844	23.7065	.001779	.0422
563	316969	23.7276	.001776	.0421
564	318096	23.7487	.001773	.0421
565	319225	23.7697	.001770	.0421
566	320356	23.7908	.001767	.0420
567	321489	23.8118	.001764	.0420
568	322624	23.8328	.001761	.0420
569	323761	23.8537	.001757	.0419
570	324900	23.8747	.001754	.0419
571	326041	23.8956	.001751	.0418
572	327184	23.9165	.001748	.0418
573	328329	23.9374	.001745	.0418
574	329476	23.9583	.001742	.0417
575	330625	23.9792	.001739	.0417
576	331776	24.0000	.001736	.0417
577	332929	24.0208	.001733	.0416
578	334084	24.0416	.001730	.0416
579	335241	24.0624	.001727	.0416
580	336400	24.0832	.001724	.0415
581	337561	24.1039	.001721	.0415
582	338724	24.1247	.001718	.0415
583	339889	24.1454	.001715	.0414
584	341056	24.1661	.001712	.0414
585	342225	24.1868	.001709	.0413

Tables

TABLE A
(*Continued*)

n	n^2	$\sqrt{n}$	$\dfrac{1}{n}$	$\dfrac{1}{\sqrt{n}}$
586	343396	24.2074	.001706	.0413
587	344569	24.2281	.001704	.0413
588	345744	24.2487	.001701	.0412
589	346921	24.2693	.001698	.0412
590	348100	24.2899	.001695	.0412
591	349281	24.3105	.001692	.0411
592	350464	24.3311	.001689	.0411
593	351649	24.3516	.001686	.0411
594	352836	24.3721	.001684	.0410
595	354025	24.3926	.001681	.0410
596	355216	24.4131	.001678	.0410
597	356409	24.4336	.001675	.0409
598	357604	24.4540	.001672	.0409
599	358801	24.4745	.001669	.0409
600	360000	24.4949	.001667	.0408
601	361201	24.5153	.001664	.0408
602	362404	24.5357	.001661	.0408
603	363609	24.5561	.001658	.0407
604	364816	24.5764	.001656	.0407
605	366025	24.5967	.001653	.0407
606	367236	24.6171	.001650	.0406
607	368449	24.6374	.001647	.0406
608	369664	24.6577	.001645	.0406
609	370881	24.6779	.001642	.0405
610	372100	24.6982	.001639	.0405
611	373321	24.7184	.001637	.0405
612	374544	24.7386	.001634	.0404
613	375769	24.7588	.001631	.0404
614	376996	24.7790	.001629	.0404
615	378225	24.7992	.001626	.0403
616	379456	24.8193	.001623	.0403
617	380689	24.8395	.001621	.0403
618	381924	24.8596	.001618	.0402
619	383161	24.8797	.001616	.0402
620	384400	24.8998	.001613	.0402
621	385641	24.9199	.001610	.0401
622	386884	24.9399	.001608	.0401
623	388129	24.9600	.001605	.0401
624	389376	24.9800	.001603	.0400
625	390625	25.0000	.001600	.0400
626	391876	25.0200	.001597	.0400
627	393129	25.0400	.001595	.0399
628	394384	25.0599	.001592	.0399
629	395641	25.0799	.001590	.0399
630	396900	25.0998	.001587	.0398
631	398161	25.1197	.001585	.0398
632	399424	25.1396	.001582	.0398
633	400689	25.1595	.001580	.0397
634	401956	25.1794	.001577	.0397
635	403225	25.1992	.001575	.0397

TABLE A
(*Continued*)

n	n^2	$\sqrt{n}$	$\dfrac{1}{n}$	$\dfrac{1}{\sqrt{n}}$
636	404496	25.2190	.001572	.0397
637	405769	25.2389	.001570	.0396
638	407044	25.2587	.001567	.0396
639	408321	25.2784	.001565	.0396
640	409600	25.2982	.001563	.0395
641	410881	25.3180	.001560	.0395
642	412164	25.3377	.001558	.0395
643	413449	25.3574	.001555	.0394
644	414736	25.3772	.001553	.0394
645	416025	25.3969	.001550	.0394
646	417316	25.4165	.001548	.0393
647	418609	25.4362	.001546	.0393
648	419904	25.4558	.001543	.0393
649	421201	25.4755	.001541	.0393
650	422500	25.4951	.001538	.0392
651	423801	25.5147	.001536	.0392
652	425104	25.5343	.001534	.0392
653	426409	25.5539	.001531	.0391
654	427716	25.5734	.001529	.0391
655	429025	25.5930	.001527	.0391
656	430336	25.6125	.001524	.0390
657	431649	25.6320	.001522	.0390
658	432964	25.6515	.001520	.0390
659	434281	25.6710	.001517	.0390
660	435600	25.6905	.001515	.0389
661	436921	25.7099	.001513	.0389
662	438244	25.7294	.001511	.0389
663	439569	25.7488	.001508	.0388
664	440896	25.7682	.001506	.0388
665	442225	25.7876	.001504	.0388
666	443556	25.8070	.001502	.0387
667	444889	25.8263	.001499	.0387
668	446224	25.8457	.001497	.0387
669	447561	25.8650	.001495	.0387
670	448900	25.8844	.001493	.0386
671	450241	25.9037	.001490	.0386
672	451584	25.9230	.001488	.0386
673	452929	25.9422	.001486	.0385
674	454276	25.9615	.001484	.0385
675	455625	25.9808	.001481	.0385
676	456976	26.0000	.001479	.0385
677	458329	26.0192	.001477	.0384
678	459684	26.0384	.001475	.0384
679	461041	26.0576	.001473	.0384
680	462400	26.0768	.001471	.0383
681	463761	26.0960	.001468	.0383
682	465124	26.1151	.001466	.0383
683	466489	26.1343	.001464	.0383
684	467856	26.1534	.001462	.0382
685	469225	26.1725	.001460	.0382

TABLE A
(*Continued*)

n	n²	$\sqrt{n}$	$\frac{1}{n}$	$\frac{1}{\sqrt{n}}$
686	470596	26.1916	.001458	.0382
687	471969	26.2107	.001456	.0382
688	473344	26.2298	.001453	.0381
689	474721	26.2488	.001451	.0381
690	476100	26.2679	.001449	.0381
691	477481	26.2869	.001447	.0380
692	478864	26.3059	.001445	.0380
693	480249	26.3249	.001443	.0380
694	481636	26.3439	.001441	.0380
695	483025	26.3629	.001439	.0379
696	484416	26.3818	.001437	.0379
697	485809	26.4008	.001435	.0379
698	487204	26.4197	.001433	.0379
699	488601	26.4386	.001431	.0378
700	490000	26.4575	.001429	.0378
701	491401	26.4764	.001427	.0378
702	492804	26.4953	.001425	.0377
703	494209	26.5141	.001422	.0377
704	495616	26.5330	.001420	.0377
705	497025	26.5518	.001418	.0377
706	498436	26.5707	.001416	.0376
707	499849	26.5895	.001414	.0376
708	501264	26.6083	.001412	.0376
709	502681	26.6271	.001410	.0376
710	504100	26.6458	.001408	.0375
711	505521	26.6646	.001406	.0375
712	506944	26.6833	.001404	.0375
713	508369	26.7021	.001403	.0375
714	509796	26.7208	.001401	.0374
715	511225	26.7395	.001399	.0374
716	512656	26.7582	.001397	.0374
717	514089	26.7769	.001395	.0373
718	515524	26.7955	.001393	.0373
719	516961	26.8142	.001391	.0373
720	518400	26.8328	.001389	.0373
721	519841	26.8514	.001387	.0372
722	521284	26.8701	.001385	.0372
723	522729	26.8887	.001383	.0372
724	524176	26.9072	.001381	.0372
725	525625	26.9258	.001379	.0371
726	527076	26.9444	.001377	.0371
727	528529	26.9629	.001376	.0371
728	529984	26.9815	.001374	.0371
729	531441	27.0000	.001372	.0370
730	532900	27.0185	.001370	.0370
731	534361	27.0370	.001368	.0370
732	535824	27.0555	.001366	.0370
733	537289	27.0740	.001364	.0369
734	538756	27.0924	.001362	.0369
735	540225	27.1109	.001361	.0369

TABLE A
(*Continued*)

n	n²	√n̄	$\frac{1}{n}$	$\frac{1}{\sqrt{n}}$
736	541696	27.1293	.001359	.0369
737	543169	27.1477	.001357	.0368
738	544644	27.1662	.001355	.0368
739	546121	27.1846	.001353	.0368
740	547600	27.2029	.001351	.0368
741	549081	27.2213	.001350	.0367
742	550564	27.2397	.001348	.0367
743	552049	27.2580	.001346	.0367
744	553536	27.2764	.001344	.0367
745	555025	27.2947	.001342	.0366
746	556516	27.3130	.001340	.0366
747	558009	27.3313	.001339	.0366
748	559504	27.3496	.001337	.0366
749	561001	27.3679	.001335	.0365
750	562500	27.3861	.001333	.0365
751	564001	27.4044	.001332	.0365
752	565504	27.4226	.001330	.0365
753	567009	27.4408	.001328	.0364
754	568516	27.4591	.001326	.0364
755	570025	27.4773	.001325	.0364
756	571536	27.4955	.001323	.0364
757	573049	27.5136	.001321	.0363
758	574564	27.5318	.001319	.0363
759	576081	27.5500	.001318	.0363
760	577600	27.5681	.001316	.0363
761	579121	27.5862	.001314	.0363
762	580644	27.6043	.001312	.0362
763	582169	27.6225	.001311	.0362
764	583696	27.6405	.001309	.0362
765	585225	27.6586	.001307	.0362
766	586756	27.6767	.001305	.0361
767	588289	27.6948	.001304	.0361
768	589824	27.7128	.001302	.0361
769	591361	27.7308	.001300	.0361
770	592900	27.7489	.001299	.0360
771	594441	27.7669	.001297	.0360
772	595984	27.7849	.001295	.0360
773	597529	27.8029	.001294	.0360
774	599076	27.8209	.001292	.0359
775	600625	27.8388	.001290	.0359
776	602176	27.8568	.001289	.0359
777	603729	27.8747	.001287	.0359
778	605284	27.8927	.001285	.0359
779	606841	27.9106	.001284	.0358
780	608400	27.9285	.001282	.0358
781	609961	27.9464	.001280	.0358
782	611524	27.9643	.001279	.0358
783	613089	27.9821	.001277	.0357
784	614656	28.0000	.001276	.0357
785	616225	28.0179	.001274	.0357

TABLE A
(*Continued*)

n	n^2	$\sqrt{n}$	$\dfrac{1}{n}$	$\dfrac{1}{\sqrt{n}}$
786	617796	28.0357	.001272	.0357
787	619369	28.0535	.001271	.0356
788	620944	28.0713	.001269	.0356
789	622521	28.0891	.001267	.0356
790	624100	28.1069	.001266	.0356
791	625681	28.1247	.001264	.0356
792	627264	28.1425	.001263	.0355
793	628849	28.1603	.001261	.0355
794	630436	28.1780	.001259	.0355
795	632025	28.1957	.001258	.0355
796	633616	28.2135	.001256	.0354
797	635209	28.2312	.001255	.0354
798	636804	28.2489	.001253	.0354
799	638401	28.2666	.001252	.0354
800	640000	28.2843	.001250	.0354
801	641601	28.3019	.001248	.0353
802	643204	28.3196	.001247	.0353
803	644809	28.3373	.001245	.0353
804	646416	28.3549	.001244	.0353
805	648025	28.3725	.001242	.0352
806	649636	28.3901	.001241	.0352
807	651249	28.4077	.001239	.0352
808	652864	28.4253	.001238	.0352
809	654481	28.4429	.001236	.0352
810	656100	28.4605	.001235	.0351
811	657721	28.4781	.001233	.0351
812	659344	28.4956	.001232	.0351
813	660969	28.5132	.001230	.0351
814	662596	28.5307	.001229	.0351
815	664225	28.5482	.001227	.0350
816	665856	28.5657	.001225	.0350
817	667489	28.5832	.001224	.0350
818	669124	28.6007	.001222	.0350
819	670761	28.6182	.001221	.0349
820	672400	28.6356	.001220	.0349
821	674041	28.6531	.001218	.0349
822	675684	28.6705	.001217	.0349
823	677329	28.6880	.001215	.0349
824	678976	28.7054	.001214	.0348
825	680625	28.7228	.001212	.0348
826	682276	28.7402	.001211	.0348
827	683929	28.7576	.001209	.0348
828	685584	28.7750	.001208	.0348
829	687241	28.7924	.001206	.0347
830	688900	28.8097	.001205	.0347
831	690561	28.8271	.001203	.0347
832	692224	28.8444	.001202	.0347
833	693889	28.8617	.001200	.0346
834	695556	28.8791	.001199	.0346
835	697225	28.8964	.001198	.0346

TABLE A
(*Continued*)

n	n^2	$\sqrt{n}$	$\dfrac{1}{n}$	$\dfrac{1}{\sqrt{n}}$
836	698896	28.9137	.001196	.0346
837	700569	28.9310	.001195	.0346
838	702244	28.9482	.001193	.0345
839	703921	28.9655	.001192	.0345
840	705600	28.9828	.001190	.0345
841	707281	29.0000	.001189	.0345
842	708964	29.0172	.001188	.0345
843	710649	29.0345	.001186	.0344
844	712336	29.0517	.001185	.0344
845	714025	29.0689	.001183	.0344
846	715716	29.0861	.001182	.0344
847	717409	29.1033	.001181	.0344
848	719104	29.1204	.001179	.0343
849	720801	29.1376	.001178	.0343
850	722500	29.1548	.001176	.0343
851	724201	29.1719	.001175	.0343
852	725904	29.1890	.001174	.0343
853	727609	29.2062	.001172	.0342
854	729316	29.2233	.001171	.0342
855	731025	29.2404	.001170	.0342
856	732736	29.2575	.001168	.0342
857	734449	29.2746	.001167	.0342
858	736164	29.2916	.001166	.0341
859	737881	29.3087	.001164	.0341
860	739600	29.3258	.001163	.0341
861	741321	29.3428	.001161	.0341
862	743044	29.3598	.001160	.0341
863	744769	29.3769	.001159	.0340
864	746496	29.3939	.001157	.0340
865	748225	29.4109	.001156	.0340
866	749956	29.4279	.001155	.0340
867	751689	29.4449	.001153	.0340
868	753424	29.4618	.001152	.0339
869	755161	29.4788	.001151	.0339
870	756900	29.4958	.001149	.0339
871	758641	29.5127	.001148	.0339
872	760384	29.5296	.001147	.0339
873	762129	29.5466	.001145	.0338
874	763876	29.5635	.001144	.0338
875	765625	29.5804	.001143	.0338
876	767376	29.5973	.001142	.0338
877	769129	29.6142	.001140	.0338
878	770884	29.6311	.001139	.0337
879	772641	29.6479	.001138	.0337
880	774400	29.6648	.001136	.0337
881	776161	29.6816	.001135	.0337
882	777924	29.6985	.001134	.0337
883	779689	29.7153	.001133	.0337
884	781456	29.7321	.001131	.0336
885	783225	29.7489	.001130	.0336

n	n^2	$\sqrt{n}$	$\dfrac{1}{n}$	$\dfrac{1}{\sqrt{n}}$
886	784996	29.7658	.001129	.0336
887	786769	29.7825	.001127	.0336
888	788544	29.7993	.001126	.0336
889	790321	29.8161	.001125	.0335
890	792100	29.8329	.001124	.0335
891	793881	29.8496	.001122	.0335
892	795664	29.8664	.001121	.0335
893	797449	29.8831	.001120	.0335
894	799236	29.8998	.001119	.0334
895	801025	29.9166	.001117	.0334
896	802816	29.9333	.001116	.0334
897	804609	29.9500	.001115	.0334
898	806404	29.9666	.001114	.0334
899	808201	29.9833	.001112	.0334
900	810000	30.0000	.001111	.0333
901	811801	30.0167	.001110	.0333
902	813604	30.0333	.001109	.0333
903	815409	30.0500	.001107	.0333
904	817216	30.0666	.001106	.0333
905	819025	30.0832	.001105	.0332
906	820836	30.0998	.001104	.0332
907	822649	30.1164	.001103	.0332
908	824464	30.1330	.001101	.0332
909	826281	30.1496	.001100	.0332
910	828100	30.1662	.001099	.0331
911	829921	30.1828	.001098	.0331
912	831744	30.1993	.001096	.0331
913	833569	30.2159	.001095	.0331
914	835396	30.2324	.001094	.0331
915	837225	30.2490	.001093	.0331
916	839056	30.2655	.001092	.0330
917	840889	30.2820	.001091	.0330
918	842724	30.2985	.001089	.0330
919	844561	30.3150	.001088	.0330
920	846400	30.3315	.001087	.0330
921	848241	30.3480	.001086	.0330
922	850084	30.3645	.001085	.0329
923	851929	30.3809	.001083	.0329
924	853776	30.3974	.001082	.0329
925	855625	30.4138	.001081	.0329
926	857476	30.4302	.001080	.0329
927	859329	30.4467	.001079	.0328
928	861184	30.4631	.001078	.0328
929	863041	30.4795	.001076	.0328
930	864900	30.4959	.001075	.0328
931	866761	30.5123	.001074	.0328
932	868624	30.5287	.001073	.0328
933	870489	30.5450	.001072	.0327
934	872356	30.5614	.001071	.0327
935	874225	30.5778	.001070	.0327

n	n^2	$\sqrt{n}$	$\dfrac{1}{n}$	$\dfrac{1}{\sqrt{n}}$
936	876096	30.5941	.001068	.0327
937	877969	30.6105	.001067	.0327
938	879844	30.6268	.001066	.0327
939	881721	30.6431	.001065	.0326
940	883600	30.6594	.001064	.0326
941	885481	30.6757	.001063	.0326
942	887364	30.6920	.001062	.0326
943	889249	30.7083	.001060	.0326
944	891136	30.7246	.001059	.0325
945	893025	30.7409	.001058	.0325
946	894916	30.7571	.001057	.0325
947	896809	30.7734	.001056	.0325
948	898704	30.7896	.001055	.0325
949	900601	30.8058	.001054	.0325
950	902500	30.8221	.001053	.0324
951	904401	30.8383	.001052	.0324
952	906304	30.8545	.001050	.0324
953	908209	30.8707	.001049	.0324
954	910116	30.8869	.001048	.0324
955	912025	30.9031	.001047	.0324
956	913936	30.9192	.001046	.0323
957	915849	30.9354	.001045	.0323
958	917764	30.9516	.001044	.0323
959	919681	30.9677	.001043	.0323
960	921600	30.9839	.001042	.0323
961	923521	31.0000	.001041	.0323
962	925444	31.0161	.001040	.0322
963	927369	31.0322	.001038	.0322
964	929296	31.0483	.001037	.0322
965	931225	31.0644	.001036	.0322
966	933156	31.0805	.001035	.0322
967	935089	31.0966	.001034	.0322
968	937024	31.1127	.001033	.0321
969	938961	31.1288	.001032	.0321
970	940900	31.1448	.001031	.0321
971	942841	31.1609	.001030	.0321
972	944784	31.1769	.001029	.0321
973	946729	31.1929	.001028	.0321
974	948676	31.2090	.001027	.0320
975	950625	31.2250	.001026	.0320
976	952576	31.2410	.001025	.0320
977	954529	31.2570	.001024	.0320
978	956484	31.2730	.001022	.0320
979	958441	31.2890	.001021	.0320
980	960400	31.3050	.001020	.0319
981	962361	31.3209	.001019	.0319
982	964324	31.3369	.001018	.0319
983	966289	31.3528	.001017	.0319
984	968256	31.3688	.001016	.0319
985	970225	31.3847	.001015	.0319

Tables

n	n^2	$\sqrt{n}$	$\dfrac{1}{n}$	$\dfrac{1}{\sqrt{n}}$
986	972196	31.4006	.001014	.0318
987	974169	31.4166	.001013	.0318
988	976144	31.4325	.001012	.0318
989	978121	31.4484	.001011	.0318
990	980100	31.4643	.001010	.0318
991	982081	31.4802	.001009	.0318
992	984064	31.4960	.001008	.0318
993	986049	31.5119	.001007	.0317
994	988036	31.5278	.001006	.0317
995	990025	31.5436	.001005	.0317
996	992016	31.5595	.001004	.0317
997	994009	31.5753	.001003	.0317
998	996004	31.5911	.001002	.0317
999	998001	31.6070	.001001	.0316
1000	1000000	31.6228	.001000	.0316

Appendixes

TABLE B
Percent of area under the normal curve between X̄ and z

z	.00	.01	.02	.03	.04	.05	.06	.07	.08	.09
0.0	00.00	00.40	00.80	01.20	01.60	01.99	02.39	02.79	03.19	03.59
0.1	03.98	04.38	04.78	05.17	05.57	05.96	06.36	06.75	07.14	07.53
0.2	07.93	08.32	08.71	09.10	09.48	09.87	10.26	10.64	11.03	11.41
0.3	11.79	12.17	12.55	12.93	13.31	13.68	14.06	14.43	14.80	15.17
0.4	15.54	15.91	16.28	16.64	17.00	17.36	17.72	18.08	18.44	18.79
0.5	19.15	19.50	19.85	20.19	20.54	20.88	21.23	21.57	21.90	22.24
0.6	22.57	22.91	23.24	23.57	23.89	24.22	24.54	24.86	25.17	25.49
0.7	25.80	26.11	26.42	26.73	27.04	27.34	27.64	27.94	28.23	28.52
0.8	28.81	29.10	29.39	29.67	29.95	30.23	30.51	30.78	31.06	31.33
0.9	31.59	31.86	32.12	32.38	32.64	32.90	33.15	33.40	33.65	33.89
1.0	34.13	34.38	34.61	34.85	35.08	35.31	35.54	35.77	35.99	36.21
1.1	36.43	36.65	36.86	37.08	37.29	37.49	37.70	37.90	38.10	38.30
1.2	38.49	38.69	38.88	39.07	39.25	39.44	39.62	39.80	39.97	40.15
1.3	40.32	40.49	40.66	40.82	40.99	41.15	41.31	41.47	41.62	41.77
1.4	41.92	42.07	42.22	42.36	42.51	42.65	42.79	42.92	43.06	43.19
1.5	43.32	43.45	43.57	43.70	43.83	43.94	44.06	44.18	44.29	44.41
1.6	44.52	44.63	44.74	44.84	44.95	45.05	45.15	45.25	45.35	45.45
1.7	45.54	45.64	45.73	45.82	45.91	45.99	46.08	46.16	46.25	46.33
1.8	46.41	46.49	46.56	46.64	46.71	46.78	46.86	46.93	46.99	47.06
1.9	47.13	47.19	47.26	47.32	47.38	47.44	47.50	47.56	47.61	47.67
2.0	47.72	47.78	47.83	47.88	47.93	47.98	48.03	48.08	48.12	48.17
2.1	48.21	48.26	48.30	48.34	48.38	48.42	48.46	48.50	48.54	48.57
2.2	48.61	48.64	48.68	48.71	48.75	48.78	48.81	48.84	48.87	48.90
2.3	48.93	48.96	48.98	49.01	49.04	49.06	49.09	49.11	49.13	49.16
2.4	49.18	49.20	49.22	49.25	49.27	49.29	49.31	49.32	49.34	49.36
2.5	49.38	49.40	49.41	49.43	49.45	49.46	49.48	49.49	49.51	49.52
2.6	49.53	49.55	49.56	49.57	49.59	49.60	49.61	49.62	49.63	49.64
2.7	49.65	49.66	49.67	49.68	49.69	49.70	49.71	49.72	49.73	49.74
2.8	49.74	49.75	49.76	49.77	49.77	49.78	49.79	49.79	49.80	49.81
2.9	49.81	49.82	49.82	49.83	49.84	49.84	49.85	49.85	49.86	49.86
3.0	49.87									
4.0	49.997									

SOURCE: Karl Pearson, *Tables for Statisticians and Biometricians,* Cambridge University Press, London, pp. 98–101, by permission of the Biometrika Trustees.

Tables

TABLE C
Values of t at the .05 and .01 levels of confidence

degrees of freedom *level of sign.*

$N_1 - N_2 - 2 =$

Not less than 30

Df $N_1 + N_2 - 2$

df	.05	.01
1	12.706	63.657
2	4.303	9.925
3	3.182	5.841
4	2.776	4.604
5	2.571	4.032
6	2.447	3.707
7	2.365	3.499
8	2.306	3.355
9	2.262	3.250
10	2.228	3.169
11	2.201	3.106
12	2.179	3.055
13	2.160	3.012
14	2.145	2.977
15	2.131	2.947
16	2.120	2.921
17	2.110	2.898
18	2.101	2.878
19	2.093	2.861
20	2.086	2.845
21	2.080	2.831
22	2.074	2.819
23	2.069	2.807
24	2.064	2.797
25	2.060	2.787
26	2.056	2.779
27	2.052	2.771
28	2.048	2.763
29	2.045	2.756
30	2.042	2.750
40	2.021	2.704
60	2.000	2.660
120	1.980	2.617
∞	1.960	2.576

SOURCE: Ronald A. Fisher and Frank Yates, *Statistical Tables for Biological, Agricultural, and Medical Research,* 4th ed, Longman Group Ltd, London (previously published by Oliver & Boyd, Edinburgh), Table III, by permission of the authors and the publisher.

MY VALUE | TABLE VALUE
3.9 | 2.16   Reject

Appendixes

TABLE D
Values of F at the .05 and
.01 confidence levels

(df for the numerator)   P = .05

df	1	2	3	4	5	6	8	12
1	161.4	199.5	215.7	224.6	230.2	234.0	238.9	243.9
2	18.51	19.00	19.16	19.25	19.30	19.33	19.37	19.41
3	10.13	9.55	9.28	9.12	9.01	8.94	8.84	8.74
4	7.71	6.94	6.59	6.39	6.26	6.16	6.04	5.91
5	6.61	5.79	5.41	5.19	5.05	4.95	4.82	4.68
6	5.99	5.14	4.76	4.53	4.39	4.28	4.15	4.00
7	5.59	4.74	4.35	4.12	3.97	3.87	3.73	3.57
8	5.32	4.46	4.07	3.84	3.69	3.58	3.44	3.28
9	5.12	4.26	3.86	3.63	3.48	3.37	3.23	3.07
10	4.96	4.10	3.71	3.48	3.33	3.22	3.07	2.91
11	4.84	3.98	3.59	3.36	3.20	3.09	2.95	2.79
12	4.75	3.88	3.49	3.26	3.11	3.00	2.85	2.69
13	4.67	3.80	3.41	3.18	3.02	2.92	2.77	2.60
14	4.60	3.74	3.34	3.11	2.96	2.85	2.70	2.53
15	4.54	3.68	3.29	3.06	2.90	2.79	2.64	2.48
16	4.49	3.63	3.24	3.01	2.85	2.74	2.59	2.42
17	4.45	3.59	3.20	2.96	2.81	2.70	2.55	2.38
18	4.41	3.55	3.16	2.93	2.77	2.66	2.51	2.34
19	4.38	3.52	3.13	2.90	2.74	2.63	2.48	2.31
20	4.35	3.49	3.10	2.87	2.71	2.60	2.45	2.28
21	4.32	3.47	3.07	2.84	2.68	2.57	2.42	2.25
22	4.30	3.44	3.05	2.82	2.66	2.55	2.40	2.23
23	4.28	3.42	3.03	2.80	2.64	2.53	2.38	2.20
24	4.26	3.40	3.01	2.78	2.62	2.51	2.36	2.18
25	4.24	3.38	2.99	2.76	2.60	2.49	2.34	2.16
26	4.22	3.37	2.98	2.74	2.59	2.47	2.32	2.15
27	4.21	3.35	2.96	2.73	2.57	2.46	2.30	2.13
28	4.20	3.34	2.95	2.71	2.56	2.44	2.29	2.12
29	4.18	3.33	2.93	2.70	2.54	2.43	2.28	2.10
30	4.17	3.32	2.92	2.69	2.53	2.42	2.27	2.09
40	4.08	3.23	2.84	2.61	2.45	2.34	2.18	2.00
60	4.00	3.15	2.76	2.52	2.37	2.25	2.10	1.92
120	3.92	3.07	2.68	2.45	2.29	2.17	2.02	1.83
∞	3.84	2.99	2.60	2.37	2.21	2.09	1.94	1.75

(df for the denominator)

SOURCE: R. A. Fisher and F. Yates, *Statistical Tables for Biological, Agricultural, and Medical Research,* 4th ed, Longman Group Ltd, London (previously published by Oliver & Boyd, Edinburgh), Table V, by permission of the authors and the publisher.

TABLE D
(*Continued*)

(df for the numerator)   P = .01

df	1	2	3	4	5	6	8	12
1	4052	4999	5403	5625	5764	5859	5981	6106
2	98.49	99.01	99.17	99.25	99.30	99.33	99.36	99.42
3	34.12	30.81	29.46	28.71	28.24	27.91	27.49	27.05
4	21.20	18.00	16.69	15.98	15.52	15.21	14.80	14.37
5	16.26	13.27	12.06	11.39	10.97	10.67	10.27	9.89
6	13.74	10.92	9.78	9.15	8.75	8.47	8.10	7.72
7	12.25	9.55	8.45	7.85	7.46	7.19	6.84	6.47
8	11.26	8.65	7.59	7.01	6.63	6.37	6.03	5.67
9	10.56	8.02	6.99	6.42	6.06	5.80	5.47	5.11
10	10.04	7.56	6.55	5.99	5.64	5.39	5.06	4.71
11	9.65	7.20	6.22	5.67	5.32	5.07	4.74	4.40
12	9.33	6.93	5.95	5.41	5.06	4.82	4.50	4.16
13	9.07	6.70	5.74	5.20	4.86	4.62	4.30	3.96
14	8.86	6.51	5.56	5.03	4.69	4.46	4.14	3.80
15	8.68	6.36	5.42	4.89	4.56	4.32	4.00	3.67
16	8.53	6.23	5.29	4.77	4.44	4.20	3.89	3.55
17	8.40	6.11	5.18	4.67	4.34	4.10	3.79	3.45
18	8.28	6.01	5.09	4.58	4.25	4.01	3.71	3.37
19	8.18	5.93	5.01	4.50	4.17	3.94	3.63	3.30
20	8.10	5.85	4.94	4.43	4.10	3.87	3.56	3.23
21	8.02	5.78	4.87	4.37	4.04	3.81	3.51	3.17
22	7.94	5.72	4.82	4.31	3.99	3.76	3.45	3.12
23	7.88	5.66	4.76	4.26	3.94	3.71	3.41	3.07
24	7.82	5.61	4.72	4.22	3.90	3.67	3.36	3.03
25	7.77	5.57	4.68	4.18	3.86	3.63	3.32	2.99
26	7.72	5.53	4.64	4.14	3.82	3.59	3.29	2.96
27	7.68	5.49	4.60	4.11	3.78	3.56	3.26	2.93
28	7.64	5.45	4.57	4.07	3.75	3.53	3.23	2.90
29	7.60	5.42	4.54	4.04	3.73	3.50	3.20	2.87
30	7.56	5.39	4.51	4.02	3.70	3.47	3.17	2.84
40	7.31	5.18	4.31	3.83	3.51	3.29	2.99	2.66
60	7.08	4.98	4.13	3.65	3.34	3.12	2.82	2.50
120	6.85	4.79	3.95	3.48	3.17	2.96	2.66	2.34
∞	6.64	4.60	3.78	3.32	3.02	2.80	2.51	2.18

(df for the denominator)

Appendixes

TABLE E
*Chi-square values at the .05
and .01 levels of confidence*

df	.05	.01
1	3.841	6.635
2	5.991	9.210
3	7.815	11.345
4	9.488	13.277
5	11.070	15.086
6	12.592	16.812
7	14.067	18.475
8	15.507	20.090
9	16.919	21.666
10	18.307	23.209
11	19.675	24.725
12	21.026	26.217
13	22.362	27.688
14	23.685	29.141
15	24.996	30.578
16	26.296	32.000
17	27.587	33.409
18	28.869	34.805
19	30.144	36.191
20	31.410	37.566
21	32.671	38.932
22	33.924	40.289
23	35.172	41.638
24	36.415	42.980
25	37.652	44.314
26	38.885	45.642
27	40.113	46.963
28	41.337	48.278
29	42.557	49.588
30	43.773	50.892

SOURCE: R. A. Fisher and F. Yates, *Statistical Tables for Biological, Agricultural, and Medical Research,* 4th ed, Longman Group Ltd, London (previously published by Oliver & Boyd, Edinburgh), Table IV, by permission of the authors and the publisher.

df	.05	.01
1	.99692	.999877
2	.95000	.990000
3	.8783	.95873
4	.8114	.91720
5	.7545	.8745
6	.7067	.8343
7	.6664	.7977
8	.6319	.7646
9	.6021	.7348
10	.5760	.7079
11	.5529	.6835
12	.5324	.6614
13	.5139	.6411
14	.4973	.6226
15	.4821	.6055
16	.4683	.5897
17	.4555	.5751
18	.4438	.5614
19	.4329	.5487
20	.4227	.5368
25	.3809	.4869
30	.3494	.4487
35	.3246	.4182
40	.3044	.3932
45	.2875	.3721
50	.2732	.3541
60	.2500	.3248
70	.2319	.3017
80	.2172	.2830
90	.2050	.2673

TABLE F
Values of r at the .05 and .01 levels of confidence

SOURCE: R. A. Fisher and F. Yates, *Statistical Tables for Biological, Agricultural, and Medical Research,* 4th ed, Longman Group Ltd, London (previously published by Oliver & Boyd, Edinburgh), Table VI, by permission of the authors and the publisher.

TABLE G
Values of r_s at the .05 and .01 levels of confidence

N	.05	.01
5	1.000	—
6	.886	1.000
7	.786	.929
8	.738	.881
9	.683	.833
10	.648	.794
12	.591	.777
14	.544	.714
16	.506	.665
18	.475	.625
20	.450	.591
22	.428	.562
24	.409	.537
26	.392	.515
28	.377	.496
30	.364	.478

SOURCE: E. G. Olds, *The Annals of Mathematical Statistics,* "Distribution of the Sum of Squares of Rank Differences for Small Numbers of Individuals," 1938, vol. 9, and "The 5 Percent Significance Levels for Sums of Squares of Rank Differences and a Correction," 1949, vol. 20, by permission of the Institute of Mathematical Statistics.

TABLE H
Random numbers

Row	1	2	3	4	5	6	7	8	9	10	11	12	13	14	15	16	17	18	19
1	9	8	9	6	9	9	0	9	6	3	2	3	3	8	6	8	4	4	2
2	3	5	6	1	7	4	1	3	2	6	8	6	0	4	7	5	2	0	3
3	4	0	6	1	6	9	6	1	5	9	5	4	5	4	8	6	7	4	0
4	6	5	6	3	1	6	8	6	7	2	0	7	2	3	2	1	5	0	9
5	2	4	9	7	9	1	0	3	9	6	7	4	1	5	4	9	6	9	8
6	7	6	1	2	7	5	6	9	4	8	4	2	8	5	2	4	1	8	0
7	8	2	1	3	4	7	4	6	3	0	7	5	0	9	2	9	0	6	1
8	6	9	5	6	5	6	0	9	0	7	7	1	4	1	8	3	1	9	3
9	7	2	1	9	9	8	0	1	6	1	6	2	3	6	9	5	5	8	4
10	2	9	0	7	3	0	8	9	6	3	3	8	5	5	6	5	2	0	9
11	9	3	5	4	5	7	4	0	3	0	1	0	4	3	3	9	5	3	2
12	9	7	5	7	9	4	8	6	8	7	6	1	6	8	2	5	5	5	3
13	4	1	7	8	6	8	1	0	5	8	8	6	1	6	8	2	9	0	4
14	5	0	8	3	3	4	5	4	4	2	5	3	0	4	9	6	1	2	3
15	3	5	0	2	9	4	1	0	0	3	9	0	5	8	6	0	9	9	6
16	0	3	8	2	3	5	1	0	1	0	6	8	5	2	4	8	0	3	8
17	1	7	2	9	1	2	7	8	4	7	0	3	3	1	5	8	2	7	3
18	5	0	5	7	9	5	8	7	8	9	3	5	3	4	4	6	1	1	3
19	7	7	3	3	5	3	6	1	3	2	8	5	4	1	4	8	3	9	0
20	1	0	9	1	3	8	2	5	3	0	3	8	0	9	3	3	0	4	5
21	1	3	8	5	1	8	5	9	4	1	9	3	9	3	6	5	9	8	4
22	8	6	4	7	8	7	5	9	4	1	9	3	9	3	6	5	9	8	4
23	0	6	9	6	5	1	0	3	2	6	7	7	4	9	6	0	3	4	0
24	7	6	7	4	7	0	8	3	8	7	3	2	5	1	2	4	2	9	7
25	3	2	3	8	1	3	1	8	7	4	5	9	0	0	2	4	1	2	1
26	9	2	1	6	4	2	3	8	7	6	2	6	2	6	4	8	1	0	1
27	3	7	4	2	2	8	1	7	8	0	6	0	0	0	3	2	2	9	7
28	0	7	8	0	8	5	1	5	2	6	5	8	7	5	3	0	5	9	6
29	7	4	2	3	3	2	6	0	0	6	5	2	2	3	6	3	9	0	4
30	1	8	2	7	5	9	5	3	6	5	2	9	9	1	1	7	3	4	3
31	4	3	1	8	7	0	6	0	8	6	5	0	1	0	4	0	6	1	5
32	8	5	8	0	6	1	4	1	2	0	4	4	1	4	7	6	3	5	1
33	4	5	8	5	0	4	5	8	3	9	2	8	7	8	9	0	8	4	3
34	5	0	2	5	4	9	2	2	1	1	0	0	5	4	8	7	6	4	0
35	0	8	1	7	0	6	3	3	4	7	6	2	6	8	9	3	4	1	4
36	2	5	9	3	4	6	0	7	5	2	0	0	9	6	0	8	2	2	5
37	2	1	3	1	3	7	8	9	8	4	9	3	8	0	2	2	1	8	1
38	3	8	8	6	8	5	1	3	3	4	6	7	2	6	3	4	8	6	7
39	0	9	9	8	5	9	8	4	4	2	2	1	1	0	1	7	6	1	3
40	2	2	3	5	3	9	7	4	4	2	1	4	0	5	8	2	3	0	8

TABLE H
(*Continued*)

| Column Number | | | | | | | | | | | | | | | | | | | | | |
20	21	22	23	24	25	26	27	28	29	30	31	32	33	34	35	36	37	38	39	40	Row
0	9	7	1	1	9	1	2	7	3	5	1	8	4	0	4	1	0	6	0	3	1
8	3	7	7	9	1	4	9	9	5	9	2	0	1	6	1	2	6	6	7	0	2
2	5	6	3	7	8	3	3	8	4	3	9	3	9	0	0	9	8	3	5	2	3
4	7	0	8	6	6	5	9	6	2	7	3	5	9	0	1	8	0	9	6	9	4
0	9	8	7	3	5	6	8	8	1	2	0	2	3	2	6	4	3	1	9	7	5
5	1	8	8	4	7	0	1	7	6	8	2	1	6	3	2	1	8	1	8	3	6
1	3	7	8	6	9	5	4	1	7	3	8	7	1	5	6	5	6	4	3	6	7
5	9	0	1	5	2	8	6	5	5	7	8	1	8	7	1	2	4	0	4	1	8
2	2	5	5	2	1	8	6	9	8	9	8	0	5	8	9	9	4	1	3	4	9
1	3	4	2	8	5	0	7	9	8	4	3	5	8	0	9	4	6	6	0	5	10
2	6	8	6	6	4	7	1	5	1	6	4	6	7	6	0	8	7	3	5	2	11
8	6	0	1	4	2	9	8	6	8	0	7	6	5	1	9	1	3	7	0	3	12
9	5	7	0	9	8	7	6	9	0	6	5	4	0	3	6	5	6	3	5	0	13
2	2	3	4	7	8	0	2	0	8	0	3	4	9	2	5	7	7	8	6	4	14
2	4	6	1	0	5	0	6	1	4	9	4	7	3	9	1	7	6	4	5	8	15
6	3	4	8	1	6	9	5	6	2	0	4	6	1	6	8	1	9	9	1	1	16
9	0	5	1	3	6	1	9	5	4	1	2	5	4	2	9	5	6	2	4	0	17
3	6	7	0	3	5	3	7	4	1	7	5	4	8	3	7	4	8	5	7	2	18
4	3	6	6	3	6	3	0	0	9	4	2	2	5	1	8	9	5	1	9	7	19
1	0	6	9	0	2	7	3	9	8	4	0	6	9	8	2	3	2	8	0	4	20
9	1	3	5	7	9	6	2	4	3	4	6	4	9	1	3	1	7	5	2	2	21
6	4	2	2	2	1	4	5	2	2	8	3	2	1	2	6	6	0	1	8	9	22
7	2	6	9	0	7	5	3	2	5	6	2	7	6	3	8	1	4	1	5	1	23
8	2	8	2	4	4	4	2	9	1	9	8	3	4	4	1	0	4	6	9	6	24
7	3	1	4	3	0	4	7	1	3	7	4	8	6	7	3	2	6	6	2	0	25
0	6	4	5	8	3	1	4	8	1	8	3	1	6	4	3	0	2	8	7	3	26
4	2	2	8	3	2	1	9	3	0	1	7	5	9	0	9	1	2	5	8	2	27
2	9	8	7	2	0	6	4	0	2	7	1	3	1	6	8	7	0	9	2	5	28
0	8	0	5	6	8	2	4	3	6	1	3	5	2	3	5	9	8	6	2	1	29
0	1	7	6	1	5	7	9	0	3	5	3	3	2	4	2	8	5	6	4	0	30
5	1	9	8	5	2	4	5	1	7	5	3	2	4	6	7	9	9	6	7	2	31
0	3	6	6	3	7	8	6	9	7	2	8	9	0	7	2	9	4	0	8	6	32
5	0	0	0	2	0	8	9	0	1	0	6	2	0	4	6	9	6	5	4	9	33
1	9	4	4	2	6	4	2	4	1	0	2	7	9	6	8	7	5	6	9	3	34
0	0	5	3	8	3	2	7	5	0	4	7	6	4	6	3	0	4	7	5	3	35
6	2	6	2	0	6	0	1	4	8	9	6	5	9	7	3	6	7	6	5	4	36
6	3	9	0	3	5	0	9	1	2	0	5	9	7	3	2	5	9	3	0	2	37
9	7	3	3	5	4	0	6	4	9	4	7	9	1	4	3	9	7	7	1	8	38
1	9	6	2	9	4	2	9	7	0	3	8	8	9	5	7	0	6	9	7	2	39
5	9	4	5	8	6	2	3	0	6	2	9	8	6	3	0	4	1	0	7	6	40

SOURCE: N. M. Downie and R. W. Heath, *Basic Statistical Methods*, 3rd ed., Harper & Row, New York, 1970. Reprinted by permission of Harper & Row.

TABLE I
Percentage points of the
studentized range

MSw df	α	$k = Number\ of\ Means$									
		2	3	4	5	6	7	8	9	10	11
5	.05	3.64	4.60	5.22	5.67	6.03	6.33	6.58	6.80	6.99	7.17
	.01	5.70	6.98	7.80	8.42	8.91	9.32	9.67	9.97	10.24	10.48
6	.05	3.46	4.34	4.90	5.30	5.63	5.90	6.12	6.32	6.49	6.65
	.01	5.24	6.33	7.03	7.56	7.97	8.32	8.61	8.87	9.10	9.30
7	.05	3.34	4.16	4.68	5.06	5.36	5.61	5.82	6.00	6.16	6.30
	.01	4.95	5.92	6.54	7.01	7.37	7.68	7.94	8.17	8.37	8.55
8	.05	3.26	4.04	4.53	4.89	5.17	5.40	5.60	5.77	5.92	6.05
	.01	4.75	5.64	6.20	6.62	6.96	7.24	7.47	7.68	7.86	8.03
9	.05	3.20	3.95	4.41	4.76	5.02	5.24	5.43	5.59	5.74	5.87
	.01	4.60	5.43	5.96	6.35	6.66	6.91	7.13	7.33	7.49	7.65
10	.05	3.15	3.88	4.33	4.65	4.91	5.12	5.30	5.46	5.60	5.72
	.01	4.48	5.27	5.77	6.14	6.43	6.67	6.87	7.05	7.21	7.36
11	.05	3.11	3.82	4.26	4.57	4.82	5.03	5.20	5.35	5.49	5.61
	.01	4.39	5.15	5.62	5.97	6.25	6.48	6.67	6.84	6.99	7.13
12	.05	3.08	3.77	4.20	4.51	4.75	4.95	5.12	5.27	5.39	5.51
	.01	4.32	5.05	5.50	5.84	6.10	6.32	6.51	6.67	6.81	6.94
13	.05	3.06	3.73	4.15	4.45	4.69	4.88	5.05	5.19	5.32	5.43
	.01	4.26	4.96	5.40	5.73	5.98	6.19	6.37	6.53	6.67	6.79
14	.05	3.03	3.70	4.11	4.41	4.64	4.83	4.99	5.13	5.25	5.36
	.01	4.21	4.89	5.32	5.63	5.88	6.08	6.26	6.41	6.54	6.66
15	.05	3.01	3.67	4.08	4.37	4.59	4.78	4.94	5.08	5.20	5.31
	.01	4.17	4.84	5.25	5.56	5.80	5.99	6.16	6.31	6.44	6.55
16	.05	3.00	3.65	4.05	4.33	4.56	4.74	4.90	5.03	5.15	5.26
	.01	4.13	4.79	5.19	5.49	5.72	5.92	6.08	6.22	6.35	6.46
17	.05	2.98	3.63	4.02	4.30	4.52	4.70	4.86	4.99	5.11	5.21
	.01	4.10	4.74	5.14	5.43	5.66	5.85	6.01	6.15	6.27	6.38
18	.05	2.97	3.61	4.00	4.28	4.49	4.67	4.82	4.96	5.07	5.17
	.01	4.07	4.70	5.09	5.38	5.60	5.79	5.94	6.08	6.20	6.31
19	.05	2.96	3.59	3.98	4.25	4.47	4.65	4.79	4.92	5.04	5.14
	.01	4.05	4.67	5.05	5.33	5.55	5.73	5.89	6.02	6.14	6.25
20	.05	2.95	3.58	3.96	4.23	4.45	4.62	4.77	4.90	5.01	5.11
	.01	4.02	4.64	5.02	5.29	5.51	5.69	5.84	5.97	6.09	6.19
24	.05	2.92	3.53	3.90	4.17	4.37	4.54	4.68	4.81	4.92	5.01
	.01	3.96	4.55	4.91	5.17	5.37	5.54	5.69	5.81	5.92	6.02
30	.05	2.89	3.49	3.85	4.10	4.30	4.46	4.60	4.72	4.82	4.92
	.01	3.89	4.45	4.80	5.05	5.24	5.40	5.54	5.65	5.76	5.85
40	.05	2.86	3.44	3.79	4.04	4.23	4.39	4.52	4.63	4.73	4.82
	.01	3.82	4.37	4.70	4.93	5.11	5.26	5.39	5.50	5.60	5.69
60	.05	2.83	3.40	3.74	3.98	4.16	4.31	4.44	4.55	4.65	4.73
	.01	3.76	4.28	4.59	4.82	4.99	5.13	5.25	5.36	5.45	5.53
120	.05	2.80	3.36	3.68	3.92	4.10	4.24	4.36	4.47	4.56	4.64
	.01	3.70	4.20	4.50	4.71	4.87	5.01	5.12	5.21	5.30	5.37
∞	.05	2.77	3.31	3.63	3.86	4.03	4.17	4.29	4.39	4.47	4.55
	.01	3.64	4.12	4.40	4.60	4.76	4.88	4.99	5.08	5.16	5.23

SOURCE: E. S. Pearson and H. O. Hartley, *Biometrika Tables for Statisticians*, Vol. 1. 3rd ed., Cambridge Press, New York, 1966, by permission of the Biometrika Trustees.

Appendix C
List of Formulas

$$\begin{aligned}\text{Percentile} \atop \text{rank}\end{aligned} = \begin{aligned}c\%\text{ below the} \\ \text{lower limit of} \\ \text{critical interval}\end{aligned}$$
$$+ \left[\frac{\text{score} - \begin{aligned}\text{lower limit of} \\ \text{critical interval}\end{aligned}}{\text{size of critical interval}} \left(\begin{aligned}\%\text{ in} \\ \text{critical} \\ \text{interval}\end{aligned} \right) \right]$$

29

$$\text{Position of median} = \frac{N + 1}{2}$$

44

$$\overline{X} = \frac{\Sigma X}{N}$$

46

$$x = X - \overline{X}$$

46

$$\overline{X} = \frac{\Sigma fX}{N}$$

47

$$\text{Median} = \begin{aligned}\text{lower limit} \\ \text{of median} \\ \text{interval}\end{aligned} + \left(\frac{\dfrac{N}{2} - \begin{aligned}\text{cf below the} \\ \text{lower limit of} \\ \text{median interval}\end{aligned}}{f \text{ in median interval}} \right) \begin{aligned}\text{size of} \\ \text{interval}\end{aligned}$$

53

$$\text{MD} = \frac{\Sigma|x|}{N}$$

61

$$\sigma = \sqrt{\frac{\Sigma x^2}{N}}$$

64

$$\sigma = \sqrt{\frac{\Sigma X^2}{N} - \overline{X}^2}$$

65

$$\sigma = \sqrt{\frac{\Sigma fX^2}{N} - \overline{X}^2}$$

66

$$z = \frac{X - \overline{X}}{\sigma}$$

88

$$X = z\sigma + \overline{X}$$

88

$$\text{Probability} = \frac{\text{number of times the outcome can occur}}{\text{total number of times any outcome can occur}}$$

89

$$z = \frac{\overline{X} - M}{\sigma_{\overline{X}}}$$

108

$$\sigma_{\overline{X}} = \frac{s}{\sqrt{N - 1}}$$

110

$$95\% \text{ confidence interval} = \overline{X} \pm (1.96)\,\sigma_{\overline{X}}$$

112

99% confidence interval $= \overline{X} \pm (2.58)\, \sigma_{\overline{X}}$ — 114

$$\sigma_P = \sqrt{\frac{P(1 - P)}{N}}$$ — 116

95% confidence interval $= P \pm (1.96)\, \sigma_P$ — 117

$$z = \frac{(\overline{X}_1 - \overline{X}_2) - 0}{\sigma_{\text{diff}}}$$ — 130

$$\sigma_{\text{diff}} = \sqrt{\sigma_{\overline{X}_1}{}^2 + \sigma_{\overline{X}_2}{}^2}$$ — 134

$$t = \frac{\overline{X}_1 - \overline{X}_2}{\sigma_{\text{diff}}}$$ — 138

$$\sigma_{\text{diff}} = \sqrt{\left(\frac{N_1 s_1{}^2 + N_2 s_2{}^2}{N_1 + N_2 - 2}\right)\left(\frac{1}{N_1} + \frac{1}{N_2}\right)}$$ — 142

$$s = \sqrt{\frac{\Sigma D^2}{N} - (\overline{X}_1 - \overline{X}_2)^2}$$ — 145

$$SS_{\text{within}} = \Sigma X_1{}^2 + \Sigma X_2{}^2 + \Sigma X_3{}^2 + \Sigma X_4{}^2$$ — 155

$$SS_{\text{bet}} = \Sigma(\overline{X} - \overline{X}_{\text{total}})^2 N$$ — 155

$$SS_{\text{total}} = SS_{\text{bet}} + SS_{\text{within}}$$ — 156

$$SS_{\text{total}} = \Sigma(X - \overline{X}_{\text{total}})^2$$ — 156

$$SS_{\text{total}} = \Sigma X^2{}_{\text{total}} - \frac{(\Sigma X_{\text{total}})^2}{N_{\text{total}}}$$ — 157

$$SS_{\text{bet}} = \left[\Sigma \frac{(\Sigma X)^2}{N}\right] - \frac{(\Sigma x_{\text{total}})^2}{N_{\text{total}}}$$ — 158

$$SS_{\text{within}} = \Sigma \left[(\Sigma X^2) - \frac{(\Sigma X)^2}{N}\right]$$ — 159

$$MS_{\text{bet}} = \frac{SS_{\text{bet}}}{df_{\text{bet}}}$$ — 159

$$MS_{\text{within}} = \frac{SS_{\text{within}}}{df_{\text{within}}}$$ — 160

$$F = \frac{MS_{\text{bet}}}{MS_{\text{within}}}$$ — 161

$$HSD = q\alpha \sqrt{\frac{MS_{\text{within}}}{n}}$$ — 165

$$\chi^2 = \Sigma \frac{(f_o - f_e)^2}{f_e}$$

172

$$\chi^2 = \frac{N(AD - BC)^2}{(A + B)(C + D)(A + C)(B + D)}$$

178

$$\chi^2 = \Sigma \frac{(|f_o - f_e| - .50)^2}{f_e}$$

179

$$\chi^2 = \frac{N(|AD - BC| - N/2)^2}{(A + B)(C + D)(A + C)(B + D)}$$

180

$$\chi_r^2 = \frac{12}{N_k(k + 1)} \Sigma (\Sigma R_i)^2 - 3N(k + 1)$$

189

$$H = \frac{12}{N(N + 1)} \Sigma \left[\frac{(\Sigma R_i)^2}{n} \right] - 3(N + 1)$$

192

$$r = \frac{\Sigma (z_X z_Y)}{N}$$

203

$$r = \frac{N \Sigma XY - (\Sigma X)(\Sigma Y)}{\sqrt{[N \Sigma X^2 - (\Sigma X)^2][N \Sigma Y^2 - (\Sigma Y)^2]}}$$

206

$$t = \frac{r \sqrt{N - 2}}{\sqrt{1 - r^2}}$$

207

$$Y' = r \left(\frac{s_y}{s_x} \right) X - r \left(\frac{s_y}{s_x} \right) \overline{X} + \overline{Y}$$

211

$$r_s = 1 - \frac{6 \Sigma D^2}{N(N^2 - 1)}$$

215

$$G = \frac{\Sigma f_a - \Sigma f_i}{\Sigma f_a + \Sigma f_i}$$

221

$$z = G \sqrt{\frac{\Sigma f_a - \Sigma f_i}{N(1 - G^2)}}$$

227

$$\phi = \sqrt{\frac{\chi^2}{N}}$$

229

$$C = \sqrt{\frac{\chi^2}{N + \chi^2}}$$

231

$$V = \sqrt{\frac{\chi^2}{N(k - 1)}}$$

232

$$\lambda = \frac{F_{iv} - M_{dv}}{N - M_{dv}}$$

233

Appendix D
Glossary

1   **Accidental sampling:** a nonrandom sampling method whereby the researcher includes the most convenient cases in his sample.
2   **Addition rule:** the probability of obtaining any one of several different outcomes equals the sum of their separate probabilities.
3   **Alpha error (type I error):** the error of rejecting the null hypothesis when we should have accepted it.
4   **Analysis of variance:** a statistical test which makes a single overall decision as to whether a significant difference is present among three or more sample means.
5   **Area under the normal curve:** that area which lies between the curve and the base line containing 100 percent or all of the cases in any given normal distribution.
6   **Bar graph (histogram):** a graphic method in which rectangular bars indicate the frequencies for the range of score values or categories.
7   **Beta error (type II error):** the error of accepting the null hypothesis when we should have rejected it.
8   **Between-groups sum of squares:** the sum of the squared deviations of every sample mean from the total mean.
9   **Bimodal distribution:** a frequency distribution containing two or more modes.
10 **Central tendency:** what is "average" or "typical" of a set of data; a value generally located toward the middle or center of a distribution.
11 **Chi square:** a nonparametric test of significance for differences

between two or more samples whereby expected frequencies are compared against obtained frequencies.

12 **Class interval:** a category in a group distribution containing more than one score value.

13 **Class limit:** the point midway between adjacent class intervals which serves to close the gap between them.

14 **Cluster sampling:** a random sampling method whereby sample members are selected on a random basis from a number of well-delineated areas known as clusters (or primary sampling units).

15 **Confidence interval:** the range of mean values (proportions) within which the true population mean (proportion) is likely to fall.

16 **Contingency coefficient:** based on chi square, a measure of the degree of association for nominal data arranged in a table larger than 2×2.

17 **Correlation:** the strength and direction of the relationship between two variables.

18 **Correlation coefficient:** generally ranging between -1.00 and $+1.00$, a number in which both the strength and direction of correlation are expressed.

19 **Cramér's V:** an alternative to the contingency coefficient which measures the degree of association for nominal data arranged in a table larger than 2×2.

20 **Cumulative frequency:** the total number of cases having any given score or a score that is lower.

21 **Cumulative frequency polygon:** a graphic method in which cumulative frequencies or cumulative percentages are depicted.

22 **Cumulative percentage:** the percent of cases having any score or a score that is lower.

23 **Curvilinear correlation:** a relationship between X and Y that begins as either positive or negative and then reverses direction.

24 **Deciles:** percentile ranks which divide the 100-unit scale by tens.

25 **Degrees of freedom:** in small sample comparisons, a statistical compensation for the failure of the sampling distribution of differences to assume the shape of the normal curve.

26 **Deviation (deviation score):** the distance and direction of any raw score from the mean.

27 **Expected frequencies:** the terms of the null hypothesis for chi square, according to which the relative frequency is expected to be the same from one group to another.

28 **F ratio:** the result of an analysis of variance, a statistical technique which indicates the size of the between-groups mean square relative to the size of the within-groups mean square.

29 **.05 level of confidence:** a level of probability at which the null hypothesis is rejected if an obtained sample difference occurs by chance only 5 times or less out of 100.

30 **Frequency distribution:** a table containing the categories, score values, or class intervals and their frequency of occurrence.

31 **Frequency polygon:** a graphic method in which frequencies are indicated by a series of points placed over the score values or midpoints of each class interval and connected with a straight line which is dropped to the base line at either end.

32 **Friedman's two-way analysis of variance by ranks:** a nonparametric alternative to the *t* ratio which is employed to compare the same sample measured twice but which requires only ordinal data.

33 **Goodman's and Kruskal's gamma:** an alternative to the rank-order correlation coefficient for measuring the degree of association between ordinal-level variables.

34 **Grouped frequency distribution:** a table which indicates the frequency of occurrence of cases located within a series of class intervals.

35 **Hypothesis:** an idea about the nature of reality which is testable by means of systematic research.

36 **Interval level of measurement:** the process of assigning a score to cases so that the magnitude of differences between them is known and meaningful.

37 **Judgment sampling (purposive sampling):** a nonrandom sampling method whereby logic, common sense, or sound judgment are used to select a sample that is representative of a larger population.

38 **Kruskal-Wallis one-way analysis of variance by ranks:** a nonparametric alternative to the analysis of variance which is employed to compare several independent samples but which requires only ordinal-level data.

39 **Kurtosis:** the peakedness of a symmetrical distribution.

40 **Lambda:** a measure of association for nominal data which indicates the degree to which we can reduce the error in predicting values of one variable from values of another.

41 **Leptokurtic:** characteristic of a symmetrical distribution which is quite peaked or tall.

42 **Level of confidence (significance level):** a level of probability at which the null hypothesis can be rejected with confidence and the research hypothesis can be accepted with confidence.

43 **Mean:** the sum of a set of scores divided by the total number of scores in the set. A measure of central tendency.

44 **Mean deviation:** the sum of the absolute deviations from the mean divided by the number of scores in a distribution. A measure of variability which indicates the average of deviations from the mean.

45 **Mean square (variance):** a measure of variation obtained by dividing between-groups sum of squares or within-groups sum of squares by the appropriate degrees of freedom.

46 **Measurement:** the use of a series of numbers in the data-analysis stage of research.

47 **Median:** the middlemost point in a frequency distribution. A measure of central tendency.

48 **Median test:** a nonparametric test of significance for determining the probability that two random samples have been drawn from populations with the same median.

49 **Mesokurtic:** characteristic of a symmetrical distribution which is neither very peaked nor very flat.

50 **Midpoint:** the middlemost score value in a class interval.

51 **Multiplication rule:** the probability of obtaining a combination of mutually exclusive outcomes equals the product of their separate probabilities.

52 **Negative correlation:** the direction of relationship wherein individuals who score high on the X variable score low on the Y variable; individuals who score low on the X variable score high on the Y variable.

53 **Negatively skewed distribution:** a distribution in which more respondents receive high than low scores, resulting in a longer tail on the left than on the right.

54 **95 percent confidence interval:** the range of mean values (proportions) within which there are 95 chances out of 100 that the true population mean (proportion) will fall.

55 **99 percent confidence interval:** the range of mean values (proportions) within which there are 99 chances out of 100 that the true population mean (proportion) will fall.

56 **Nominal level of measurement:** the process of placing cases into categories and counting their frequency of occurrence.

57 **Nonparametric test:** a statistical procedure which makes no assumptions about the way the characteristic being studied is distributed in the population and requires only ordinal or nominal data.

58 **Nonrandom sampling:** a sampling method whereby each and every population member does not have an equal chance of being drawn into the sample.

59 **Normal curve:** a smooth, symmetrical distribution which is bell-shaped and unimodal.

60 **Null hypothesis:** the hyothesis which explains any observed difference between samples as being a chance occurrence resulting from sampling error alone.

61 **Obtained frequencies:** in a chi-square analysis the results that are actually obtained when conducting a study.

62 **.01 level of confidence:** a level of probability at which the null hypothesis is rejected if an obtained sample difference occurs by chance only 1 time or less out of 100.

63 **Ordinal level of measurement:** the process of ordering or ranking cases in terms of the degree to which they have any given characteristic.

64 **Parametric test:** a statistical procedure which requires that the

characteristic studied be normally distributed in the population and that the researcher have interval data.

65 **Pearson correlation coefficient (r):** a correlation coefficient for interval data which is equal to the mean of the z-score products for the *X* and *Y* variables.

66 **Percentage:** a method of standardizing for size which indicates the frequency of occurrence of a category per 100 cases.

67 **Percentile rank:** a single number which indicates the percent of cases in a distribution falling below any given score.

68 **Phi coefficient:** based on chi square, a measure of the degree of association for nominal data arranged in a 2 × 2 table.

69 **Pie chart:** a circular graph whose pieces add up to 100 percent.

70 **Platykurtic:** characteristic of a symmetrical distribution which is rather flat.

71 **Population (universe):** any set of individuals who share at least one characteristic.

72 **Positive correlation:** the direction of a relationship wherein individuals who score high on the *X* variable also score high on the *Y* variable; individuals who score low on the *X* variable also score low on the *Y* variable.

73 **Positively skewed distribution:** a distribution in which more respondents receive low than high scores, resulting in a longer tail on the right than on the left.

74 **Power:** the ability of a statistical test to reject the null hypothesis when it is actually false and should be rejected.

75 **Primary sampling unit (cluster):** in cluster sampling a well-delineated area considered to include characteristics found in the entire population.

76 **Probability:** the relative frequency of occurrence of an event or outcome. The number of times any given event could occur out of 100.

77 **Proportion:** a method for standardizing for size which compares the number of cases in any given category with the total number of cases in the distribution.

78 **Quartiles:** percentile ranks which divide the 100-unit scale by twenty-fives.

79 **Quota sampling:** a nonrandom sampling method whereby diverse characteristics of a population are sampled in the proportions they occupy in the population.

80 **Random sampling:** a sampling method whereby each and every population member has an equal chance of being drawn into the sample.

81 **Range:** the difference between the highest and lowest scores in a distribution. A measure of variability.

82 **Rate:** a kind of ratio which indicates a comparison between the number of actual cases and the number of potential cases.

83 **Ratio:** a method of standardizing for size which compares the

number of cases falling into one category with the number of cases falling into another category.

84 **Regression analysis:** a technique employed in predicting values of one variable (Y) from knowledge of values of another variable (X).

85 **Regression line:** a straight line drawn through the scatter diagram which represents the best possible "fit" for making predictions from X to Y.

86 **Research hypothesis:** the hypothesis which regards any observed difference between samples as reflecting a true population difference and not just sampling error.

87 **Sample:** a smaller number of individuals taken from some population (for the purpose of generalizing to the entire population from which it was taken).

88 **Sampling distribution of differences:** a frequency distribution of a large number of differences between random sample means that have been drawn from a given population.

89 **Sampling distribution of means:** a frequency distribution of a large number of random sample means which have been drawn from the same population.

90 **Sampling error:** the inevitable difference between a random sample and its population based on chance and chance alone.

91 **Scatter diagram:** a graph that shows the way scores on any two variables X and Y are scattered throughout the range of possible score values.

92 **Simple random sampling:** a random sampling method whereby a table of random numbers is employed to select a sample that is representative of a larger population.

93 **Size:** the number of score values contained in a class interval.

94 **Skewness:** departure from symmetry.

95 **Spearman's rank-order (r_s):** a correlation coefficient for data that have been ranked or ordered with respect to the presence of a given characteristic.

96 **Standard deviation:** the square root of the mean of the squared deviations from the mean of a distribution. A measure of variability which indicates the average of deviations from the mean.

97 **Standard error of the difference:** an estimate of the standard deviation of the sampling distribution of differences based on the standard deviations of two random samples.

98 **Standard error of the mean:** an estimate of the standard deviation of the sampling distribution of means based on the standard deviation of a single random sample.

99 **Standard error of the proportion:** an estimate of the standard deviation of the sampling distribution of proportions based on the proportion obtained in a single random sample.

100 **Statistically significant difference:** a sample difference that reflects a real population difference and not just a sampling error.

101 **Straight-line correlation:** either a positive or negative correlation,

so that the points in a scatter diagram tend to form a straight line through the center of the graph.

102 **Strata:** the homogeneous subgroups employed for purposes of stratified sampling.

103 **Stratified sampling:** a random sampling method whereby the population is first divided into more homogeneous subgroups from which simple random samples are then drawn.

104 **Strength of correlation:** degree of association between two variables.

105 **Sum of squares:** the sum of squared deviations from a mean.

106 **Systematic sampling:** a random sampling method whereby every nth member of a population is included in the sample.

107 *t* **ratio:** for small sample comparisons, a statistical technique that indicates the direction and degree that a sample mean difference falls from zero on a scale of standard deviation units.

108 **Total sum of squares:** the sum of the squared deviations of every raw score from the total mean of the study.

109 **Tukey's HSD (honestly significant difference):** a procedure for the multiple comparison of means after a significant F ratio has been obtained.

110 **Unimodal distribution:** a frequency distribution containing a single mode.

111 **Variability:** the manner in which the scores are scattered around the center of the distribution. Also known as dispersion or spread.

112 **Variable:** any characteristic which varies from one individual to another. Hypotheses usually contain an independent variable (cause) and a dependent variable (effect).

113 **Within-groups sum of squares:** the sum of the squared deviations of every raw score from its sample mean.

114 **Yates's correction:** in the chi-square analysis, factor for small expected frequencies which reduces the overestimate of the chi-square value and yields a more conservative result.

115 *z* **score (standard score):** a value which indicates the direction and degree that any given raw score deviates from the mean of a distribution on a scale of standard deviation units.

116 *z* **score for sample mean differences:** a value which indicates the direction and degree that any given sample mean difference falls from zero (the mean of the sampling distribution of differences) on a scale of standard deviation units.

117 **Zero correlation:** no relationship between the x and y variables.

References

Anderson, Theodore R., and Morris Zelditch, Jr., *A Basic Course in Statistics,* Holt, Rinehart and Winston, New York, 1968.

Blalock, Hubert M., *Social Statistics,* McGraw-Hill, New York, 1960.

Campbell, Stephen K., *Flaws and Fallacies in Statistical Thinking,* Prentice-Hall, Englewood Cliffs, N.J., 1974.

Champion, Dean J., *Basic Statistics for Social Research,* Chandler, San Francisco, 1970.

Chase, Clinton I., *Elementary Statistical Procedures,* McGraw-Hill, New York, 1967.

Cohen, Lillian, *Statistical Methods for Social Scientists,* Prentice-Hall, Englewood Cliffs, N.J., 1954.

Courts, Frederick A., *Psychological Statistics,* The Dorsey Press, Homewood, Ill., 1966.

Dixon, Wilfrid J., and Frank J. Massey, *Introduction to Statistical Analysis,* McGraw-Hill, New York, 1957.

Dornbusch, Sanford M., and Calvin F. Schmid, *A Primer of Social Statistics,* McGraw-Hill, New York, 1955.

Downey, Kenneth J., *Elementary Social Statistics,* Random House, New York, 1975.

Downie, Norville M., and R. W. Heath, *Basic Statistical Methods,* Harper & Row, New York, 1974.

Edwards, A. L., *Experimental Design in Psychological Research,* Holt, Rinehart and Winston, New York, 1960.

Edwards, Allen L., *Statistical Methods for the Behavioral Sciences,* Holt, Rinehart and Winston, New York, 1967.

Ferguson, George A., *Statistical Analysis in Psychology and Education,* McGraw-Hill, New York, 1966.

Freeman, Linton C., *Elementary Applied Statistics,* Wiley, New York, 1965.

Freund, John E., *Modern Elementary Statistics,* Prentice-Hall, Englewood Cliffs, N.J., 1960.

Fried, Robert, *Introduction to Statistics,* Oxford University, 1969.

Guilford, Jay P., *Fundamental Statistics in Psychology and Education,* McGraw-Hill, New York, 1956.

Hagood, Margaret J., and Daniel O. Price, *Statistics for Sociologists,* Holt, Rinehart and Winston, New York, 1952.

Hammond, Kenneth R., and James E. Householder, *Introduction to the Statistical Method,* Knopf, New York, 1963.

Huff, Darrell, *How to Lie with Statistics,* Wiley, New York, 1966.

Loether, Herman J., and Donald G. McTavish, *Inferential Statistics for Sociologists,* Allyn & Bacon, Boston, 1974.

McNemar, Quinn, *Psychological Statistics,* Wiley, New York, 1962.

Meyers, Lawrence S., and Neal E. Grossen, *Behavioral Research,* Freeman, San Francisco, 1974.

Mueller, John H., Karl F. Schuessler, and Herbert L. Costner, *Statistical Reasoning in Sociology,* Houghton Mifflin, Boston, 1970.

Palumbo, Dennis J., *Statistics in Political and Behavioral Science,* Appleton, New York, 1969.

Popham, W. James, and Kenneth A. Sirotnik, *Educational Statistics,* Harper & Row, New York, 1973.

Runyon, Richard P., and Audrey Haber, *Fundamentals of Behavioral Statistics,* Addison-Wesley, Reading, Mass., 1971.

Siegel, Sidney, *Nonparametric Statistics for the Behavioral Sciences,* McGraw-Hill, New York, 1956.

Spence, Janet T., Benton J. Underwood, Carl P. Duncan, and John W. Cotton, *Elementary Statistics,* Appleton, New York, 1968.

Walker, Helen Mary, and Joseph Lev, *Elementary Statistical Methods,* Holt, Rinehart and Winston, New York, 1958.

Wallis, Wilson A., and Harry Roberts, *The Nature of Statistics,* Free Press, New York, 1965.

Welkowitz, Joan, Robert B. Ewen, and Jacob Cohen, *Introductory Statistics for the Behavioral Sciences,* Academic, New York, 1971.

Williams, Frederick, *Reasoning with Statistics,* Holt, Rinehart and Winston, New York, 1968.

Winer, B. J., *Statistical Principles in Experimental Design,* McGraw-Hill, New York, 1962.

Answers to Selected Problems

Chapter 1

1. nominal
2. ordinal
3. interval
4. ordinal
5. interval
6. nominal
7. ordinal
8. nominal
9. interval (assuming equal intervals between points on scale)

Chapter 2

1. (a) 51%, (b) 27%, (c) $P = .51$, (d) $P = .27$
2. (a) 71%, (b) 74%, (c) $P = .71$, (d) $P = .74$
3. $\frac{4}{24} = \frac{1}{6}$
4. 156.25
5. $\frac{15}{20} = \frac{3}{4}$
6. There are 85.71 live births per every 1000 women of childbearing age.
7. 66.67%

8.
Class Interval	f
10–12	11
7–9	16
4–6	9
1–3	4
	$N = 40$

a. 3
b. 9.5–12.5
   6.5–9.5
   3.5–6.5
   .5–3.5
c. 11
   8
   5
   2
d. *cf*
   40
   29
   13
   4
e. *c%*
   100
   72.5
   32.5
   10.0

9. Class Interval	f
99–100	10
97–98	16
95–96	22
93–94	11
91–92	5
	$N = 65$

a. 2
b. 98.5–100.5
   96.5–98.5
   94.5–96.5
   92.5–94.5
   90.5–92.5
c. 99.5
   97.5
   95.5
   93.5
   91.5
d. *cf*
   64
   54
   38
   16
   5
e. *c%*
   100
   84.4
   59.4
   25.0
   7.8

10. (a) 59.38, (b) 14.59
11. (a) 84.82, (b) 29.64

Chapter 4

1. (a) 9, (b) 6, (c) 5.71
2. (a) 9 and 1, (b) 5, (c) 5.13
3. (a) 5, (b) 5, (c) 32.71
4. (a) 1, (b) 2.5, (c) 3
5. (a) 10, (b) 10, (c) 9.63
6. (a) 3 and 6, (b) 4, (c) 4.1
7. (a) 8, (b) 8, (c) 7.67
8. (a) 6, (b) 4.5, (c) 4.17
9. (a) 4, (b) 5, (c) 6
10. (a) 12, (b) 7, (c) 7.86
11. (a) 0, (b) $+12.5$, (c) -5.5, (d) $+.5$
12. (a) $+1.0$, (b) $-.5$, (c) $+3.3$, (d) 0
13. (a) 7.5, (b) -12, (c) 0, (d) -4.5
14. (a) 4, (b) 4, (c) 4.13
15. (a) 3, (b) 3, (c) 3.19
16. (a) 6, (b) 6, (c) 6.26
17. (a) 12, (b) 12.3, (c) 12.79
18. (a) 84.5, (b) 82.4, (c) 80.39
19. (a) 12, (b) 11.76, (c) 12

Chapter 5

1. (a) 6, (b) 1.92, (c) 2.15
2. (a) Class A = 5, Class B = 3, (b) Class A = 1.67, Class B = .83, (c) Class A = 1.89, Class B = .96
3. (a) 4, (b) 1.28, (c) 1.50
4. 2.70
5. 1.6
6. 1.19
7. 1.54
8. 1.40
9. (a) 49, (b) 10.51, (c) 12.46
10. (a) 14, (b) 2.47, (c) 3.25
11. (a) 19, (b) 3.71, (c) 4.66

Chapter 6

1. (a) 68.26%, (b) 95.44%, (c) 99.74%
2. (a) $+.38$, (b) -1.15, (c) -1.69, (d) $+2.08$, (e) 0, (f) .77, (g) $+2.69$
3. (a) $-.75$, (b) $+.18$, (c) $+.96$, (d) -1.96, (e) $+1.61$, (f) $+.36$, (g) $-.54$
4. (a) 5.37%, (b) $P = .05$, (c) 7.14%, (d) $P = .07$, (e) $P = .43$, (f) $P = .86$ (g) $P = .18$
5. (a) .38%, (b) P is less than .01, (c) 40.82%, (d) $P = .41$, (e) 25.14%, (f) $P = .25$ (g) $P = .18$ (h) $P = .0007$

Chapter 7

1. .27
2. (a) 2.40 $\longleftrightarrow$ 3.46, (b) 2.23 $\longleftrightarrow$ 3.63
3. .35
4. (a) 5.10 $\longleftrightarrow$ 6.48, (b) 4.89 $\longleftrightarrow$ 6.69
5. .39
6. (a) 4.24 $\longleftrightarrow$ 5.76, (b) 3.99 $\longleftrightarrow$ 6.01
7. (a) .07, (b) .43 $\longleftrightarrow$ .71

8. (a) .04, (b) .24 $\longleftrightarrow$ .40
9. (a) .03, (b) .19 $\longleftrightarrow$ .31

Chapter 8

1. $z = 2.50$, $P = .01$, reject the null hypothesis at .05
2. $t = 1.47$, df $= 6$, accept the null hypothesis at .05
3. $t = 1.84$, df $= 12$, accept the null hypothesis at .05
4. $t = 2.03$, df $= 16$, accept the null hypothesis at .05
5. $t = 4.31$, df $= 8$, reject the null hypothesis at .05
6. $t = .67$, df $= 8$, accept the null hypothesis at .05
7. $t = 3.90$, df $= 13$, reject the null hypothesis at .05
8. $t = 4.32$, df $= 10$, reject the null hypothesis at .05
9. $t = 2.51$, df $= 10$, reject the null hypothesis at .05
10. $t = 3.12$, df $= 5$, reject the null hypothesis at .05
11. $t = 3.85$, df $= 3$, reject the null hypothesis at .05
12. $t = 6.0$, df $= 4$, reject the null hypothesis at .05

Chapter 9

1. $F = 2.71$, df $= \frac{3}{12}$, accept the null hypothesis at .05
2. $F = 46.33$, df $= \frac{2}{8}$, reject the null hypothesis at .05
3. $F = 6.99$, df $= \frac{2}{12}$, reject the null hypothesis at .05
4. $F = 4.23$, df $= \frac{2}{12}$, reject the null hypothesis at .05
5. HSD $= 2.11$. Therefore, only $\overline{X}_1 - \overline{X}_3$ is statistically significant.
6. $F = 8.16$, df $= \frac{3}{20}$, reject the null hypothesis at .05
7. HSD $= 1.98$. Therefore, $\overline{X}_1 = \overline{X}_2, \overline{X}_1 - \overline{X}_3$, and $\overline{X}_1 - \overline{X}_4$ are statistically significant

Chapter 10

1. $\chi^2 = 1.36$, df $= 1$, accept the null hypothesis at .05
2. $\chi^2 = 8.29$, df $= 1$, reject the null hypothesis at .05
3. $\chi^2 = 2.17$, df $= 1$, accept the null hypothesis at .05
4. $\chi^2 = 1.50$, df $= 1$, accept the null hypothesis at .05
5. $\chi^2 = 1.78$, df $= 1$, accept the null hypothesis at .05
6. $\chi^2 = 17.77$, df $= 4$, reject the null hypothesis at .05
7. $\chi^2 = 17.75$, df $= 3$, reject the null hypothesis at .05
8. $\chi^2 = 2.24$, df $= 2$, accept the null hypothesis at .05
9. Mdn $= 5$, $\chi^2 = 2.07$, df $= 1$, accept the null hypothesis at .05
10. Mdn $= 6$, $\chi^2 = 19.57$, df $= 1$, reject the null hypothesis at .05
11. $\chi r^2 = 1.96$, df $= 1$, accept the null hypothesis at .05
12. $\chi r^2 = 10.20$, df $= 2$, reject the null hypothesis at .05
13. $H = 1.97$, df $= 2$, accept the null hypothesis at .05
14. $H = 10.64$, df $= 2$, reject the null hypothesis at .05

Chapter 11

1. $r = +.85$, df $= 4$, significant at .05
2. $r = -.64$, df $= 2$, not significant at .05
3. $r = +.89$, df $= 3$, significant at .05
4. $r = +.93$, df $= 3$, significant at .05
5. $r = -.95$, df $= 5$, significant at .05
6. $Y' = .52X + 1.01$; (a) $Y' = 3.61$, (b) $Y' = 2.05$, (c) $Y' = 5.69$
7. $Y' = -.99X + 10.66$; (a) $Y' = .76$, (b) $Y' = 8.68$
8. $r_s = -.53$, $N = 5$, not significant at .05
9. $r_s = -.65$, $N = 8$, not significant at .05
10. $r_s = -.89$, $N = 7$, significant at .05

11. $r_s = -.80$, $N = 5$, not significant at .05
12. $G = +.60$, $z = .82$, not significant at .05
13. $G = -.39$, $z = 1.15$, not significant at .05
14. $\phi = .37$
15. $\phi = .17$
16. $\phi = .17$
17. (a) $C = .26$, (b) $V = .20$
18. (a) $C = .36$, (b) $V = .39$
19. (a) $C = .27$, (b) $V = .20$
20. (a) .07, (b) .08, college major has slightly greater predictive ability.

Index